纤维增强树脂基复合材料（FRP）检修教程

主　编　李璋琪　付成龙
副主编　黄频波　张　凯
　　　　喻兴亮　王稳升
主　审　倪泽军　李　铮

北京理工大学出版社
BEIJING INSTITUTE OF TECHNOLOGY PRESS

内 容 提 要

本书依据飞机结构修理专业人才培养方案编写，共6个任务，分别为损伤评估与记录、维修方案制定、去除损伤与型面准备、损伤修补、封装固化与修理后处理、技能综合训练，针对航空复合材料维修行业需求与岗位职业能力要求，以波音、空客飞机复合材料典型结构修理案例为指引，涵盖典型检修案例的工艺规范与施工标准，培养学生具备从航空复合材料制备工艺到检测与修理规范的核心能力。

本书可作为飞机结构修理和复合材料相关专业教材，也可供复合材料相关专业工程技术人员学习及复合材料修理相关岗位培训参考使用。

版权专有　侵权必究

图书在版编目（CIP）数据

纤维增强树脂基复合材料（FRP）检修教程：英、汉/李璋琪，付成龙主编. -- 北京：北京理工大学出版社，2023.8

ISBN 978-7-5763-2843-1

Ⅰ.①纤…　Ⅱ.①李…②付…　Ⅲ.①碳纤维增强复合材料－检修－高等学校－教材－英、汉　Ⅳ.①TB334

中国国家版本馆CIP数据核字（2023）第167565号

责任编辑：阎少华　　　　文案编辑：阎少华
责任校对：周瑞红　　　　责任印制：王美丽

出版发行 / 北京理工大学出版社有限责任公司
社　　址 / 北京市丰台区四合庄路6号
邮　　编 / 100070
电　　话 / （010）68914026（教材售后服务热线）
　　　　　（010）68944437（课件资源服务热线）
网　　址 / http://www.bitpress.com.cn
版 印 次 / 2023年8月第1版第1次印刷
印　　刷 / 河北鑫彩博图印刷有限公司
开　　本 / 787 mm×1092 mm　1/16
印　　张 / 12.5
字　　数 / 244千字
定　　价 / 79.00元

前　言

　　树脂基复合材料是航空航天领域重要结构材料，其具有高强度、轻量化等性能优势，在航空器飞机机翼、机身蒙皮等构件广泛应用已有数十年历史，复合材料检修也逐渐成为保证航空器持续适航性的重要内容。目前，纤维增强树脂基复合材料（FRP）结构件的修理维护的标准施工已成体系，因此对于从事复合材料检修相关任务的专业技术技能人才的培养也相应提出更高的规范标准。

　　基于"岗课赛证"四位一体教学理念，本书针对行业需求与岗位职业能力要求，根据飞机结构修理专业人才培养方案，以典型结构修理案例为载体，结合世界技能大赛飞机维修项目复合材料检修模块技术文件与核心能力要求，引用部分机型修理手册如 SRM 结构修理手册（SRM）和无损检测（NDT）手册等标准施工程序与图示，以双语活页式形式编写，致力于贴近实际修理工作情境并提供标准规范的施工程序，符合二十大报告中提出的"推进职普融通、产教融合、科教融汇，优化职业教育类型定位""加快建设国家战略人才力量，努力培养造就更多大师、战略科学家、一流科技领军人才和创新团队、青年科技人才、卓越工程师、大国工匠、高技能人才"，以及"岗课赛证"中要求的"针对实际工作需求，提升员工技能水平"的目标。其内容涵括 FRP 损伤结构维修的全部流程，详解复合材料检修案例实施与维修工卡的相关填写规范，精确把握职业素养等综合能力，满足各类型学员学习需求，可供飞机结构修理和复合材料相关专业使用，也可作为世界技能大赛训练资料，或复合材料修理相关岗位培训资料参考，引导学生在知行合一、学以致用上下功夫，增长知识、锤炼品格，符合"岗课赛证"中"贴近实际，实战性强"的要求。

　　本书以 FRP 典型损伤修理工艺为主要内容，按照通用修理工序分为六个任务：损伤评估与记录、维修方案制定、去除损伤与型面准备、损伤修补、封装固化与修理后处理和技能综合训练。本书每章内容分为六部分，包括任务情境、任务解析、任务分解、背景知识、工单和实训任务。要求学员掌握基本的文件填写与手册查询能力，具备从航空复合材料制备工艺到检测与修理规范的核心能力，并根据训练工卡不断精进岗位能力。

本书由成都航空职业技术学院李璋琪、付成龙担任主编，由成都航空职业技术学院黄频波、航空工业成都飞机工业（集团）有限责任公司张凯、成都航空职业技术学院喻兴亮、四川航空股份有限公司王稳升担任副主编，由北京飞机维修工程有限公司成都维修基地倪泽军和北京维修基地李铮主审。具体编写分工如下：李璋琪负责任务1、任务3的编写及全书英文翻译工作，付成龙编写任务5，黄频波、喻兴亮编写任务2，张凯编写任务4，王稳升编写任务6。通过本书学习，读者能够获得对航空复合材料检修岗位完整且深入的认知。本书在编著过程中参考了其他相关文献资料，未能一一注明，在此一并感谢。

由于编者水平有限，书中难免存在疏漏之处，恳请读者批评指正。

编　者

Preface

Resin-based composite materials are important structural materials in the aerospace field. They possess advantages such as high strength and lightweight properties. They have been widely used in components such as aircraft wings and fuselage skins for several decades. Composite material inspection and maintenance has gradually become an important aspect in ensuring the continued airworthiness of aircraft. Currently, standard construction practices for the repair and maintenance of fiber-reinforced resin-based composite (FRP) structures have been established. Therefore, higher standards and specifications are required for the training of professional technical personnel engaged in composite material inspection and maintenance tasks.

Based on the "job-course-competition-certificate" system, this book targets industry demand and job competency requirements. It is based on the professional talent development plan for aircraft structural repair and uses typical structural repair cases as carriers, combined with the technical documents and core competency requirements of the composite material inspection and repair module of the World Skills Competition aircraft maintenance project. The book also includes standard construction procedures and diagrams from repair manuals for certain aircraft models such as the Structural Repair Manual (SRM) and Non-destructive Testing (NDT) Manual. The book is written in bilingual loose-leaf form, aiming to be close to actual repair work scenarios and to provide standard and regulated construction procedures. It is in line with the "strengthen vocational education, promote industry-education integration, cultivate high-skilled talents" proposed in China's 20 Major Development and Planning, as well as the goal of "improving employee skills level in response to actual job requirements" required by the "job-position-competition-certificate" system.

The content of the book includes the entire process of FRP damage structural repair, detailing the relevant filling specifications of the composite material inspection and repair work card and the implementation of repair cases. It accurately grasps comprehensive abilities such as professional ethics and meets the learning needs of various types of students. It can be used by aircraft structural repair and composite material-related professionals, as well as a training reference material for the WorldSkills Competition or composite material repair–related job training, guiding students to work hard in the integration of theory and practice, and to gain knowledge and improve their character, in line with the requirements of "closeness to reality and strong practicality" in the "job–position–competition–certificate" system.

This book primarily focuses on the typical repair processes for FRP composite material damages. It is divided into six tasks based on general repair procedures: Damage Assessment and Documentation, Repair Plan Formulation, Damage Removal and Surface Preparation, Damage Repair, Encapsulation and Curing, Post–Repair Processing, and Comprehensive Skill Training. Each chapter of the book comprises six sections: Task Context, Task Analysis, Task Breakdown, Background Knowledge, Work Order, and Practical Training Task. The book emphasizes the acquisition of basic skills in document completion and manual referencing, and aims to develop core competencies ranging from aviation composite material preparation techniques to inspection and repair specifications. It encourages learners to continuously enhance their proficiency in job–related tasks by utilizing training work cards.

The book is edited by Li Zhangqi and Fu Chenglong from Chengdu Aviation Polytechnics, with Huang Pinbo from Chengdu Aviation Polytechnics, Zhang Kai from AVIC Chengdu Aircraft Industrial (Group) Co., LTD., Yu Xingliang from Chengdu Aviation Polytechnics, and Wang Wensheng from Sichuan Airlines Co., LTD as deputy editors., and reviewed by Ni Zejun of Chengdu Maintenance base and Li Zheng of Beijing Maintenance base from Beijing Aircraft Maintenance Engineering Co., LTD. The Task assignment is as follows: Li Zhangqi is responsible for the English translation and Tasks 1 and 3, Fu Chenglong for Task 5, Huang Pinbo and Yu Xingliang for Task 2, Zhang Kai for Task 4 and Wang Wensheng for Task 6. The editors sincerely hope that readers can gain a complete and in–depth understanding of the aviation composite material repair position through this book. During the writing process, the book referenced other author's relevant literature and materials, and we would like to express our gratitude to them all.

Regrettably, the editor's proficiency is limited, and as such, the book may contain omissions, inadequacies, and errors. It is humbly requested that readers provide constructive criticism and corrections should any such issues be encountered.

目 录 Contents

01

任务 1 损伤评估与记录
Task 1 Damage Assessment and Recording

子任务 1 目视检查与敲击检测·································· 4
Subtask 1 Visual Check and Tapping Test············· 4
子任务 2 FRP 超声检测······································ 15
Subtask 2 Ultrasonic Test for FRP Damages ······· 15
子任务 3 记录损伤信息······································ 39
Subtask 3 Damage Information Report ················ 39

02

任务 2 维修方案制定
Task 2 Maintenance Scheme Determination

子任务 1 SRM 查询·· 48
Subtask 1 SRM Consultation····························· 48
子任务 2 ADL 查询·· 59
Subtask 2 ADL Consultation····························· 59

03

任务 3 去除损伤与型面准备
Task 3 Damage Removal & Surface Preparation

子任务 1 清洁与干燥·· 69
Subtask 1 Cleanup and Absorbed Moisture
Removal ·· 69
子任务 2 去除损伤与打磨修理型面······················ 85
Subtask 2 Damage Removal and Surface
Preparation ··································· 85

04

任务 4 损伤修补
Task 4 Damage Repair

子任务 1 湿铺层的制作与铺贴·················· 102
Subtask 1 Wet-layup Repair Patch Preparation
and Installation ·················· 102
子任务 2 蜂窝修理芯塞安装················116
Subtask 2 Honeycomb Core Repair Plug
Installation ················116

05

任务 5 封装、固化与修理后处理
Task 5 Encapsulation, Curing and Post-repair Treatment

子任务 1 封装和预压实·················· 127
Subtask 1 Encapsulation and Debulking·············· 127
子任务 2 固化················· 140
Subtask 2 Curing ················· 140
子任务 3 修理后处理················· 150
Subtask 3 Post-repair Treatment ················· 150

06

任务 6 技能综合训练
Task 6 Comprehensive Skills Training

综合训练题一 层合板单面修理 ·················· 160
综合训练题二 夹芯结构单面修理 ·················· 162
综合训练题三 B737 升降舵修理 ·················· 168
综合训练题四 夹芯结构维修方案制定与实施·············· 172

参考文献 / 191

损伤评估与记录
Damage Assessment and Recording

【任务情境 Task Scenario】

您是 WS 航空公司的航线大修人员，并持有复合材料部件维修授权书。早晨你到岗后，查阅某型号飞机工作记录，显示：夜班维修人员已经航后检查，机身存在两处凹痕。这两个凹痕位于机身右侧，在 FR 24~25，右翼 12~13。你的工作是对损伤进行详细检查，完成损伤评估并将检查结果填写在维护计划单中。

You are a line maintenance technician for WS Airlines, and you hold a composite material component repair authorization. When you arrive at your station in the morning, you check the work records for a specific aircraft model, which indicate that the night shift has already conducted a post–flight inspection. During this inspection, two dents were found on the right side of the fuselage: one between FR 24−25 and another between the right wing 12−13. Your task is to conduct a detailed examination of the damages, complete a damage assessment, and record the inspection results on the maintenance work order.

【任务解析 Task Analysis】

判定损伤是进行复合材料结构损伤维修工作的第一步。完成本项目训练后，应实现以下目标：

Damage assessment is the first step of damage repair for composite structures. Students should be able to realize:

知识目标 Knowledge Objectives

（1）熟悉复合材料损伤类型与机理。

Familiarize yourself with types and mechanisms of composite material damage.

（2）熟悉复合材料目视检查、敲击检测、超声检测基本原理与适用范围。

Familiarize yourself with the basic principles and scope of composite material visual check, impact detection, and ultrasonic detection.

（3）知道航线检查单、损伤报告等航线资料。

Know flight line checklists, damage reports and other flight line information.

 能力目标 **Ability Objectives**

（1）运用目视检查与敲击检测方法，准确判定复合材料损伤类型。

Use visual check and impact inspection methods to accurately identify composite material damage.

（2）使用 SRM 和 NDT 手册等行业标准，准确并规范地描述损伤位置、范围大小、深度等信息，填写结构损伤报告。

Use industry standards such as SRM and NDT manuals to accurately and consistently describe damage location, size, and depth, and fill out damage reports.

（3）确定最小修理范围以及修理计划。

Determine minimum repair scope and repair plan.

素质目标 **Emotion Objectives**

（1）机务作风与安全意识：航空安全为第一，掌握复合材料检查中的安全事项，保护自身和他人安全，同时养成严谨的机务作风，时刻保障飞机安全质量。

Maintenance culture and safety awareness: Aviation safety comes first, mastering safety issues in composite material inspections, protecting oneself and others, and developing a rigorous maintenance culture to always ensure the safety and quality of aircraft.

（2）团队合作与沟通能力：学会与团队成员配合，共同完成检修任务。培养与客户、同事、监察有效沟通的能力。

Teamwork and communication skills: Learn to collaborate with team members to complete maintenance tasks together. Develop effective communication skills with clients, colleagues, and supervisors.

正确填写结构损伤报告
Complete structural damage report correctly

损伤评估与记录
Damage assessment and recording

规范执行
目视检查/敲击检测/超声检测
Standard procedure for visual
check,tapping inspection and
ultrasonic inspection

判定损伤类型
确定修理方案
Determine damage type
Determine repair plan

子任务 1　目视检查与敲击检测
Subtask 1　Visual Check and Tapping Test

【子任务解析 Subtask Analysis】

结构损伤是指发生影响部件结构完整性的损伤，通常需要进行检测，确定损坏范围并报告损伤情况。基于复合材料的层合结构，缺陷或损伤的类型通常繁杂，在工程实践中，需采用两种或两种以上不同的检测方法对缺陷与损伤进行检测，以便互相补充和验证。

Structural damage refers to damage that affects the integrity of a component's structure, which usually requires inspection to determine the extent of the damage and report the damage situation. Based on the layered structure of composite materials, the types of defects or damages are usually complex. In engineering practice, two or more different inspection methods are required to detect defects and damages, so that they can complement and validate each other.

通过本任务的训练，需要掌握航线维修工作中常用的目视检查与敲击检测方法，掌握检测的规范操作流程，以实现鉴别常见层合板以及夹芯结构的损伤形式。

Through this training task, you are required to master the commonly used visual inspection and tap testing methods in route maintenance work, understand the standard operating procedures for inspections, and be able to identify common damage forms of composite laminates and sandwich structures.

【子任务分解 Subtask Break-down】

（1）掌握目视检查与敲击检测的规范操作流程；

Master the standard operation procedure of visual check and tapping test.

（2）理解并鉴别凹坑、分层、脱粘、划伤等常见结构损伤形式；

Understand and distinguish common classification of structural damage such as dents, delamination, debonding and scratches.

（3）查阅并理解 AC43-13-1B 与 SRM 文件。

Read and basically understand AC43-13-1B and SRM files.

小任务 1　检测任务的意义是什么？
Subtask 1　What is the point of the detection task?

先进航空复合材料具有极好的疲劳寿命和承受极限载荷的能力，因此根据维护计划文件（MPD）规定，复合材料检测任务主要针对意外损伤进行排查。MPD 涉及的检查由特定的无损检测（NDT）程序涵盖，复合材料发生意外损坏后，由 SRM 提供规范检查技术，由 NDT 管理检查报告，常规的检查方式包括绕机目视检查、小区域详细目视检查、重要结构检查。

Advanced aviation composites possess excellent fatigue life and the ability to withstand extreme load conditions. Therefore, according to the Maintenance Planning Document (MPD) guidelines, composite material inspections primarily focus on detecting unexpected damages. The inspections specified in the MPD are covered by specific Non-Destructive Testing (NDT) procedures known as NDT. In the event of accidental damage to composite materials, the Structural Repair Manual (SRM) provides the prescribed inspection techniques, while the NDT team manages the inspection reports. Routine inspection methods include visual inspections around the aircraft, detailed visual inspections of localized areas, and critical structure inspections.

航空安全始终是最重要的考虑因素，在进行复合材料检查时更是如此。为了保证自身和他人的安全，维修人员必须熟知和遵守与复合材料检查相关的安全事项，例如，合适的个人保护装备的佩戴、安全环境的维护等。此外，养成严谨的机务作风也是保障飞机安全的重要措施之一，维修人员应时刻关注细节，注重质量，严格执行工作流程和标准，以确保飞机的安全和可靠性。

Aviation safety is always the paramount consideration, especially when conducting composite material inspections. To ensure the safety of themselves and others, maintenance personnel must be familiar with and adhere to safety procedures related to composite material inspections. This includes wearing appropriate personal protective equipment and maintaining a safe working environment. Furthermore, cultivating a rigorous maintenance culture is also crucial in safeguarding aircraft safety. Maintenance personnel should always pay attention to details, prioritize quality, and strictly follow work procedures and standards to ensure the safety and reliability of the aircraft.

小任务 2 航空复合材料的损伤形式主要有哪些？
Subtask 2 What are the main damage types of aviation composites?

航空复合材料的损伤形式主要包括分层、脱粘、划伤、凿伤、凹坑、腐蚀损伤，其形貌特征如表 1-1-1 所示。

The damage forms of aerospace composite materials mainly include delamination, debonding, scratching, gouging, dents and corrosion damage. Their morphological characteristics are shown in Table 1-1-1.

表 1-1-1 航空复合材料的主要损伤形式
Table 1-1-1 Main Damage Types of Aviation Composites

损伤图示 Damage Illustrations	损伤类型 Damage Types	损伤机理 Damage Mechanism
	分层 Delamination	低速冲击损伤 Low speed impact damage
	脱粘 Debonding	湿热环境老化 Hygrothermal effect 蜂窝芯进水 Honeycomb core water intrusion
	划伤 Scratch 凿伤 Gouge	外来尖锐物机械作用 Mechanical action of foreign sharp objects
	凹坑 Dent	外来物冲击损伤 凹坑附近分层 Impacted by a foreign object Delamination around dent
	腐蚀 Corrosion	湿热环境老化 Hygrothermal effect 渗透破坏 Seepage failures

（1）分层 / 脱粘 Delamination/Debonding
由局部低速冲击损伤造成。损伤来源包括石头撞击、工具掉落、雷击、鸟击、冰雹等，均可导致分层损伤或开裂。

Localized low-speed impact damage is typically caused by sources such as rocks, dropped tools, lightning strikes, bird strikes, hail, etc., and can result in delamination or cracking damage.

由于吸湿或其他环境老化而对粘接界面造成冲击损伤或降解的结果。针对层合板等整体构件，脱粘可能发生在与周围结构的粘接界面上，如肋与蒙皮界面；针对夹层结构，蜂窝芯与内部和外部表皮之间可能发生脱粘。

The result of impact damage or degradation to the bonding interface caused by moisture absorption or other environmental aging. For integral components such as composite laminates, delamination may occur at the bonding interface with the surrounding structure, such as between the ribs and the skin. For sandwich structures, delamination may occur between the honeycomb core and the internal and external skins.

（2）划伤 / 凿伤 Scratch/Gouge

由外来尖锐物机械作用损伤蒙皮或漆层造成。划伤是指任何深度和长度的线型损伤，凿伤与划痕损伤原理一致，而凿伤损伤缺口更宽且更深。飞机在飞行过程中，机身增压载荷使机身蒙皮承受环向拉伸作用。在这种循环载荷作用下，蒙皮表面划伤会在蒙皮搭接部位逐渐形成裂纹，严重影响飞机的安全，所以要高度重视。

Foreign sharp objects can cause mechanical damage to the skin or paint. Scratches refer to any linear damage of any depth and length. The principle of gouge damage is consistent with that of scratch damage, but the gouge damage gap is wider and deeper. During flight, the pressurized load of the aircraft body causes the skin of the aircraft body to bear a circumferential tensile force. Under this cyclic load, scratches on the skin surface will gradually form cracks at the skin overlap, which seriously affects the safety of the aircraft, so it should be highly valued.

（3）凹坑 Dent

受外来物撞击造成局部凹陷，如鸟击，常见于蒙皮、壁板和整流罩等气动外形。凹坑会增加飞行阻力，且凹坑处的二次弯曲应力很大，导致疲劳强度下降，因此凹坑损伤周围通常伴有脱粘或分层损伤。

Local dents caused by external impact, such as bird strikes, are common on aerodynamic surfaces such as skin, bulkheads, and nacelles. Dents increase drag and the secondary bending stress around them is high, leading to reduced fatigue strength. Therefore, delamination or layering damage is often found around the dent area.

（4）腐蚀 Corrosion

腐蚀元件的复合材料修复应完全或部分恢复其性能或刚度。腐蚀按腐蚀形式可分为以

下几类:

The bonded composite repair of corroded elements should fully or partially restore the capacity and/or stiffness. Corrosion can be classified into the following categories based on its form:

1）全面腐蚀 Comprehensive Corrosion

也称均匀腐蚀，胶层在整个表面均匀流失。一般来说，维修包括在特殊考虑下更换原有的所有结构。

Also known as uniform corrosion, where the adhesive layer uniformly erodes over the entire surface. Typically, repairs include replacing all existing structures under special considerations.

2) 点状腐蚀 Pitting Corrosion

局部腐蚀的一种形式，是表面局部孔洞或凹坑的腐蚀。首先应该估计点蚀的强度。通常，点蚀修复包括更新板材，通过焊接形成坑，或应用塑料填充化合物。

A form of local corrosion that involves localized pitting or depression corrosion on the surface. The strength of the pitting should be evaluated first. Usually, repair of pitting includes replacement of the plate, forming a depression by welding, or application of plastic filler compounds.

3) 开槽腐蚀 Grooving Corrosion

另一种形式的局部腐蚀，通常是机械连接接头附近的局部材料损耗。其特征是出现沟槽状或裂隙状腐蚀损伤。腐蚀修复需要特别考虑损坏的部件可能需要更换。

Grooving is a another form of local corrosion that usually occurs in the local material near mechanical connection joints. It is characterized by the appearance of groove-like or fissure-like corrosion damage. Corrosion repair needs to be considered specially, and damaged parts may need to be replaced.

4) 边缘腐蚀 Edge Corrosion

结构件的局部材料损耗。通常可根据腐蚀严重程度增补加强件。

Local material loss at the free edge of structural components. Typically, reinforcement can be added depending on the severity of corrosion.

小任务 3　航空复合材料检修操作流程参考的规章文件主要有哪些?

Subtask 3　What are the main rules and documents for reference to the maintenance procedure of aviation composite inspection and repair?

（1）NTM 51-10-03 CFRP 和 GFRP 复合材料部件检测

NTM 51-10-03 Inspection of CFRP and GFRP composite component

（2）NTM 51-10-06 意外损伤一碳纤维整体结构检验的通用无损检测程序

NTM 51-10-06 Accidental damage-general NDT Procedure for the inspection of carbon fiber monolithic structure

（3）NTM 51-10-09 碳纤维结构的通用目视检查程序

NTM 51-10-09 General visual inspection procedure for carbon fiber structures

（4）NTM 51-10-10 X 射线检测以检测蜂窝夹层部件中的水分

NTM 51-10-10 X-Ray inspection to detect water in honeycomb sandwich parts

（5）NTM 51-10-19 FRP 复合材料部件的检测，使用啄木鸟检测蜂窝夹层部件

NTM 51-10-19 Inspection of FRP composite components honeycomb sandwich parts with woodpecker

小任务 4　目视检查与敲击检测的检测对象是哪些？
Subtask 4　What are the objects of visual check and tapping test?

（1）目视检查主要检查损伤位置、损伤面积和损伤程度。例如，查看有无划伤、凹坑、开裂、蒙皮凸起或其他表面缺陷。

The visual check mainly examines the location, area and extent of the damage such as scratches, dents, cracks, skin bulge, or other surficial defects.

（2）敲击检查确定脱粘和分层的范围。

Tapping test detects and determines the extent of debonding and delamination.

（3）用敲击锤等敲击工具或设备敲击蜂窝结构的蒙皮，根据不同的声响来判断蜂窝结构是否脱粘，通常适用于检测层数在三层以内的结构件分层损伤。

Tapping test hammer or other tools are used to tap the skin of the honeycomb structure, and the debonding of the honeycomb structure is judged according to different sounds. This method is usually used to detect delamination damage of laminates within three layers.

（4）若敲击回声清脆，则结构完整性良好；若敲击回声沉闷，则存在脱粘或分层损伤。结构完整性损伤导致刚度下降，振动频率下降，声音就会沉闷。

If the tapping echo is clear, the structural integrity is good; if the tapping echo is dull, there is debonding or delamination damage. Damage to the structural integrity results in a decrease in stiffness and vibration frequency, causing the sound to be dull.

工单 Work Order

耗材 / 工具 / 设备清单

Materials/Tools/Equipment List

耗材 / 工具 / 设备 Materials/Tools/Equipment	图示 Illustration
警示牌 Warning placard 防护服 Protective clothing 劳保鞋 Labor protection shoes	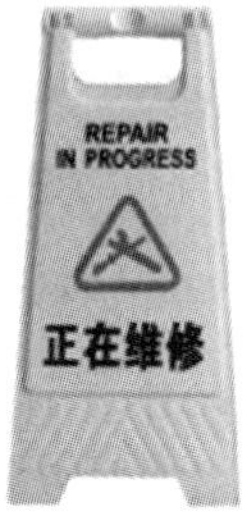 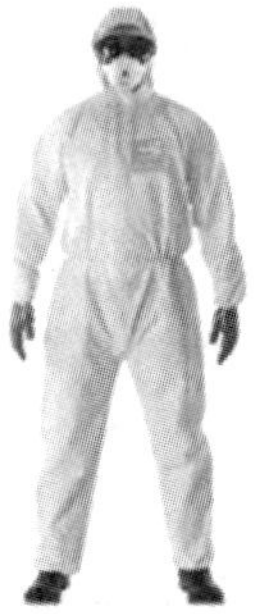 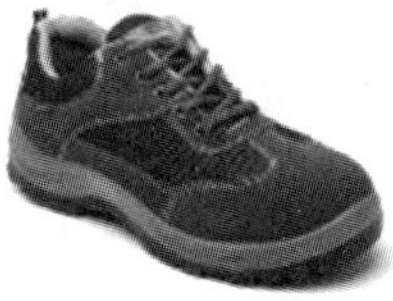
清洁工具 Cleaning tools： SOPM 规定清洁剂 SOPM specified detergent 砂纸 Sandpaper	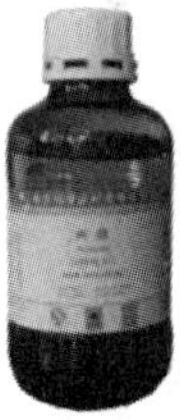 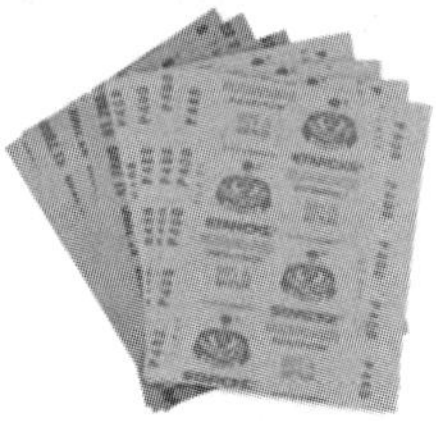
检测和绘图工具 Testing and drawing tools： 敲击锤 Tapping hammer 记号笔 Marker	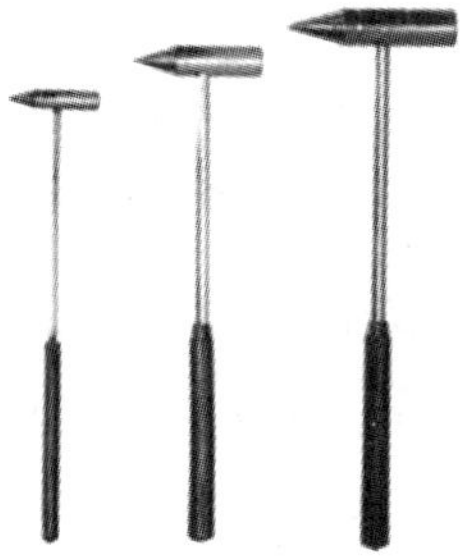 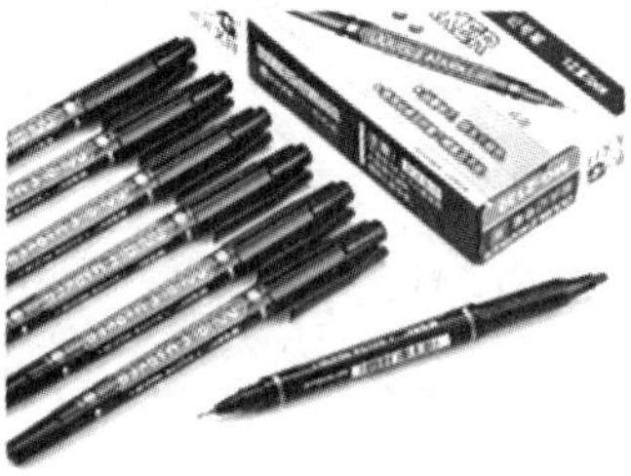

任务工单
Task Card

步骤 Steps	规章 / 指令 Regulation/Instruction
	任务名称 Task Topic 目视检查与敲击检测 Visual Inspection and Tapping Test
1. 查阅 Review	查阅检测件图纸，了解内部结构。 Review structural configuration of test parts, acknowledge of internal structure.
2. 清洁 Cleaning-up	按照 SOPM 规定，确保检查区域表面干净和光滑。 Refer to SOPM, make sure that the surface of the inspection area is clean and smooth.
3. 目视检查 Visual Check	按照 NTM 规定定位检测区域，目视检查表面损伤，填写损伤记录。 In accordance with NTM regulations, position the inspection area, visually examine the surface for damages, and fill out the damage records.
4. 敲击检测 Tapping Test	敲击路径参考图 1-1-1。 Tapping path refer to FIGURE 1-1-1. 敲击方法参考图 1-1-2。 Tapping method refer to FIGURE 1-1-2. 使用敲击测试工具以大约 10 mm（0.39 in）间距的网格模式敲击检查区域的整个表面，手持敲击锤方式参考图 1-1-3。 Use a tapping test tool to tap the entire surface of the inspection area in a grid pattern with approximately 10 mm (0.39 in) spacing, tap test hammer holding method refer to Fig. 1-1-3. 注意：必须控制敲击力度，轻且稳定。 Caution: The tap action must be light but firm action. 使用毛毡尖笔在部件表面标记存在分层 / 脱粘迹象的区域。记录所有分层 / 脱粘的位置和区域信息并形成报告。 Use a felt-tip pen to mark any delamination/debonding indication on the surface of the component. Record all damage information of the location and area and complete the detection report. 注意：蒙皮厚度和内部结构单元的变化会产生不同的声音响应，可能与分层 / 脱粘指示混淆。 Notes: Variations in skin thickness and internal structural will produce different sound responses which may be confusing with the delamination/debonding indication.

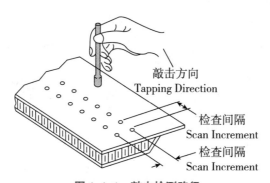

图 1-1-1 敲击检测路径
Fig.1-1-1 Tap Test Pattern

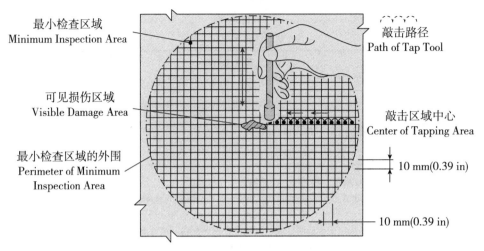

最小检查区域
Minimum Inspection Area

敲击路径
Path of Tap Tool

可见损伤区域
Visible Damage Area

敲击区域中心
Center of Tapping Area

最小检查区域的外围
Perimeter of Minimum
Inspection Area

10 mm(0.39 in)

10 mm(0.39 in)

图 1-1-2　冲击损伤敲击检测方法

Fig.1-1-2　Tapping Test in Area Surrounding Visible Impact Damage

敲击

图 1-1-3　手持敲击锤操作示意图

Fig.1-1-3　Tap Test Hammer Holding Method

目视检查与敲击检测练习题任务工卡

Visual Check and Tapping Test Exercise Task Card

（1）飞机复合材料的目视检查中常见的缺陷类型有哪些？它们的特征和影响是什么？

What are the common types of defects in visual inspection of aircraft composite materials? What are their characteristics and effects?

（2）目视检查中需要采取哪些步骤来确保检查的准确性和可靠性？

What steps need to be taken in visual inspection to ensure accuracy and reliability?

（3）目视检查与其他检测方法（如 NDT）相比有什么优劣之处？

What are the advantages and disadvantages of visual inspection compared to other testing methods (such as NDT)?

目视检查与敲击检测操作实践工序卡
Visual Inspection and Tapping Test Operation Log

1. 目视检查损伤记录：

Visual inspection of damage recorded：

2. 敲击检测损伤记录：

Inspection by tapping for damage recorded：

3. 损伤报告记录补充或 N/A：

Supplement or N/A for damage report records：

4. 维修计划记录：

Maintenance plan records：

5. 超声检查损伤记录或 N/A：

Ultrasonic inspection of damage recorded or N/A：

6. 清点归还工具：

Inventory check and return of tools：

操作总结：

Summary：

子任务 2　FRP 超声检测
Subtask 2　Ultrasonic Test for FRP Damages

【子任务解析 Subtask Analysis】

纤维增强树脂基复合材料（FRP）超声检测是一种非破坏性检测技术，利用超声波在材料内部传播时所发生的声波反射、衍射、折射等物理现象，精确检测 FRP 结构中的缺陷、损伤和异物等。

FRP Ultrasonic Testing is a non-destructive testing technique that utilizes the physical phenomena of sound wave reflection, diffraction, and refraction that occur when ultrasonic waves propagate through materials. It precisely detects defects, damages, and foreign objects within the structure of composite materials like FRP (Fiber Reinforced Plastic).

通过本子任务的训练，可以了解超声扫描的基本原理，掌握针对层合板和蜂窝夹层板复合材料的超声探伤标准施工程序。

Through the training of this subtask, you can learn the basic principles of ultrasound scanning and master the standard construction procedures for ultrasonic testing of laminated and honeycomb sandwich composite materials.

【子任务分解 Subtask Break-down】

（1）掌握超声检查层合板 FRP 的规范操作流程。

Standardize operation procedure of ultrasonic wrought plate material.

（2）掌握层合板、蜂窝夹层板 FRP 常见结构损伤的超声探伤波形特征。

Master the ultrasonic testing waveform characteristics of common structural damage in FRP laminates and honeycomb sandwich structures.

小任务 1　超声检测的对象有哪些?
Subtask 1　What are the objects of ultrasonic test?

超声检测可评估复合材料结构完整性和可靠性,其适用于各种类型的复合材料结构探伤,如碳纤维增强聚合物(CFRP)、玻璃纤维增强聚合物(GFRP)、芳香族聚酰亚胺纤维增强聚合物(PAI)、环氧树脂(EP)等,可主要检测以下类型的损伤:

Ultrasonic testing can assess the integrity and reliability of composite structures, and it is suitable for various types of composite structure probing, such as carbon fiber reinforced polymer (CFRP), glass fiber reinforced polymer (GFRP), polyaromatic imide fiber reinforced polymer (PAI), epoxy resin (EP), etc. It can mainly detect the following types of damage:

(1)分层或裂纹:超声波能够探测到复合材料中的分层或裂纹,包括纵向、横向和45度角裂纹。

Delamination or cracking: Ultrasonic waves can detect delamination or cracking in composite materials, including longitudinal, transverse, and 45-degree angle cracks.

(2)疲劳损伤:复合材料在受到循环载荷作用时容易出现疲劳损伤,超声波能够检测到这些损伤。

Fatigue damage: Composite materials are prone to fatigue damage when subjected to cyclic loading, and ultrasonic waves can detect such damage.

(3)孔隙:复合材料制造过程或修理过程中可能会发生局部贫胶或富胶导致孔隙产生,超声波能够检测到这些孔隙。

Porosity: Porosity may occur in the manufacturing or repair process of composite materials due to local resin deficiency or excess, and ultrasonic waves can detect such pores.

与目视检查和敲击检测相比,超声检测能够探测到微小的缺陷和损伤,对材料的损伤进行准确的定量和定位,通过提供材料内部缺陷的三维信息和图像,确定缺陷的形状、大小和位置,同时确定缺陷的深度和方向,也可同时借助计算机软件进行自动化处理和分析,提高检测效率和准确性。

Compared to visual inspection and tap testing, ultrasonic inspection can detect small defects and damage in materials, accurately quantify and locate the damage, provide three-dimensional information and images of internal defects in materials, determine the shape, size, and location of the defects, as well as the depth and direction of the defects. With the assistance of computer software, it can also perform automated processing and analysis, improving inspection efficiency and accuracy.

小任务 2　超声波的分类有哪些?
Subtask 2　What are the classifications of ultrasound?

能引起听觉的频率范围为 20 Hz~20 kHz 的声波称为可闻声波，低于 20 Hz 频率的声波称为次声波，高于 20 kHz 频率的声波称为超声波。在超声检测系统中，常用电脉冲激励探头的压电晶片使其产生机械振动，这种振动在与其接触的介质中传播形成超声波。

Sound waves with frequencies that can cause auditory perception, ranging from 20 Hz to 20 kHz, are called audible sound waves. Sound waves with frequencies below 20 Hz are called infrasound waves, while those with frequencies above 20 kHz are called ultrasound waves. In ultrasonic detection systems, piezoelectric crystals in the electric pulse−excited probes are commonly used to generate mechanical vibrations that propagate in contact with the probe to form ultrasound waves.

超声波的波形是以波动中介质质点振动方向与波传播方向之间的关系来确定的。在超声波检测中主要应用的波形有纵波、横波、表面波（瑞利波）。

The waveform of an ultrasound wave is determined by the relationship between the vibration direction of the medium particles in the wave and the direction of wave propagation. The main waveforms used in ultrasonic detection are longitudinal waves, transverse waves, and surface waves (Rayleigh waves).

（1）纵波 L Longitudinal Wave L

介质中质点的振动方向平行于波的传播方向的一种波形称为纵波，用 L 表示，如图 1-2-1 所示。纵波中介质质点受到交变拉压应力作用并产生伸缩形变，故纵波亦称为压缩波。由于纵波中的质点疏密相间，故又称为疏密波。凡能承受拉伸或压缩应力的介质都能传播纵波。固体介质能承受拉伸或压缩应力，因此固体介质可以传播纵波。液体和气体虽然不能承受拉伸应力，但能承受应力产生的体积变化，因此液体和气体介质也可以传播纵波。

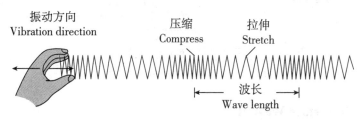

图 1-2-1　纵波示意图

Fig.1-2-1　Longitudinal wave diagram

A type of wave in which the vibration direction of the medium particles is parallel to the direction of wave propagation is called a longitudinal wave, represented by L, as shown in Fig.1-2-1. In a longitudinal wave, the medium particles experience alternating tensile and compressive stresses, resulting in elongation and contraction deformations, so it is also called a compression wave. Since the particles in a longitudinal wave are spaced in a pattern of alternating high and low density, it is also called a density wave. Any medium that can withstand tensile or compressive stress can propagate longitudinal waves. Solid media can withstand tensile or compressive stress, so they can propagate longitudinal waves. Although liquids and gases cannot withstand tensile stress, they can withstand the volume change caused by stress, so they can also propagate longitudinal waves.

（2）横波 S Transverse Wave S

介质中质点的振动方向与波的传播方向互相垂直的波形称为横波，用 S 或 T 表示，如图 1-2-2 所示。横波中介质质点受到交变的剪切应力作用并产生切变形变，故横波又称为切变波或剪切波。只有固体介质才能承受剪切应力，液体和气体介质不能承受剪切应力，故横波只能在固体介质中传播，不能在气、液介质中传播。

A type of wave in which the vibration direction of the medium particles is perpendicular to the direction of wave propagation is called a transverse wave, represented by S or T, as shown in Fig.1-2-2. In a transverse wave, the medium particles experience alternating shear stresses, resulting in shearing deformations, so it is also called a shear wave or transverse shear wave. Only solid media can withstand shear stress, while liquid and gas media cannot withstand shear stress, so transverse waves can only propagate in solid media and cannot propagate in liquid and gas media.

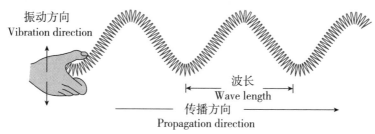

图 1-2-2　横波示意图

Fig.1-2-2　Transverse wave diagram

（3）表面波 R Surface Wave R

当介质表面受到交变应力作用时，产生沿介质表面传播的波，称为表面波，常用 R 表示，如图 1-2-3 所示。表面波是瑞利在 1887 年首先提出来的，因此又称为表面波瑞利波。

When the surface of a medium is subjected to alternating stress, a wave propagating along the surface of the medium is generated, called a surface wave, usually represented by R, as shown in Fig.1-2-3. Surface waves were first proposed by Rayleigh in 1887, so they are also called Rayleigh waves.

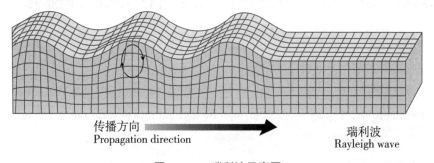

传播方向
Propagation direction

瑞利波
Rayleigh wave

图 1-2-3　瑞利波示意图
Fig.1-2-3　Rayleigh wave diagram

表面波在介质表面传播时，介质表面质点做椭圆运动，椭圆长轴垂直于波的传播方向，短轴平行于波的传播方向。椭圆运动可视为纵向振动与横向振动的合成，即纵波与横波的合成。因此表面波同横波一样只能在固体介质中传播，不能在液体或介质中传播。表面波的能量随传播深度的增加而迅速减弱。当传播深度超过两倍波长时，质点的振幅就已经很小了。因此，表面波检测只能检测到距离工件表面一倍波长深度范围内的缺陷。

When a surface wave propagates on the surface of a medium, the surface particles of the medium move in an elliptical motion, with the long axis of the ellipse perpendicular to the direction of wave propagation and the short axis parallel to the direction of wave propagation. The elliptical motion can be seen as the combination of longitudinal vibration and transverse vibration, that is, the combination of longitudinal waves and transverse waves. Therefore, like transverse waves, surface waves can only propagate in solid media and cannot propagate in liquids or gases. The energy of surface waves rapidly decreases with increasing depth of propagation. When the propagation depth exceeds twice the wavelength, the amplitude of the particle vibration is already very small. Therefore, surface wave detection can only detect defects within a depth range of one wavelength from the surface of the workpiece.

思考：根据波形定义，分析水波属于哪种波形？

Thinking: According to definition of wave types, which category of wave type does water wave belong to?

小任务 3　超声 A、B 和 C 型显示的定义
Subtask 3　Definition of Ultrasound A, B and C-Scan Displays

脉冲波超声检测仪是超声检测技术中应用最广泛的仪器。按图形显示方式，可以分为 A 型显示、B 型显示和 C 型显示几种类型。

Pulsed-wave ultrasonic flaw detectors are the most widely used instruments in ultrasonic testing. According to the display mode, they can be divided into several types, including A-scan, B-scan, and C-scan displays.

（1）A 型显示 A-scan

A 型显示又称 A 扫描或 A 超，是目前脉冲反射式探伤仪应用最广、最基本的一种类型。主要利用超声波的反射特性，在荧光屏上以纵坐标代表反射回波的声压幅度，以横坐标代表超声波的传播时间。根据反射回波在扫描基线上的刻度可以知道缺陷的位置，根据反射回波在垂直范围的刻度可以知道声压的当量大小，如图 1-2-4 所示。

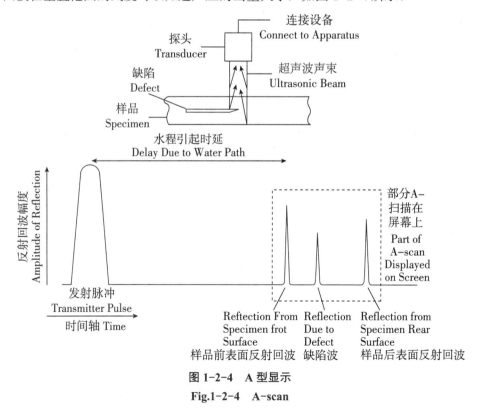

图 1-2-4　A 型显示

Fig.1-2-4　A-scan

A-made display also known as A-scan or A-ultrasound, is the most widely used and basic type of pulse-echo ultrasonic detector. It mainly utilizes the reflection characteristics of ultrasonic waves, with the vertical coordinate representing the sound pressure amplitude of the reflected echo on the screen, and the horizontal coordinate representing the propagation time of the ultrasonic wave. The position of the defect can be determined based on the scale of the reflected echo on the scanning baseline, and the equivalent sound pressure level can be determined based on the scale of the reflected echo in the vertical range, as shown in Fig.1-2-4.

（2）B 型显示 B-scan

B 型显示又称 B 扫描或 B 超，它以反射回波作为辉度调制信号，用亮点显示接收信号，在屏幕上纵坐标代表声波的传播时间，横坐标代表探头的水平位置，反映缺陷的水平延伸情况。B 型显示能直观地显示缺陷在横截面上的二维特征，获得断面直观图，如图 1-2-5 所示。

B-mode display also known as B-scan or B-ultrasound, uses reflected echoes as brightness modulation signals, displaying received signals as bright dots on the screen. The vertical axis represents the propagation time of the sound wave, and the horizontal axis represents the horizontal position of the probe, reflecting the horizontal extension of the defect. B-mode display can intuitively display the two-dimensional characteristics of defects in the cross-section, obtaining a visual sectional image, as shown in Fig.1-2-5.

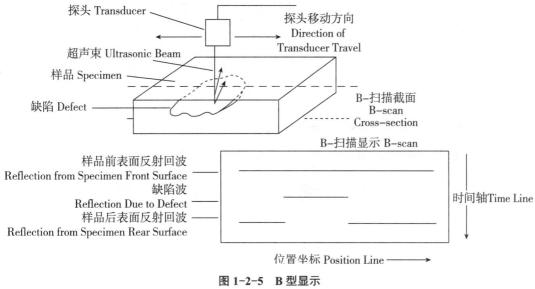

图 1-2-5　B 型显示

Fig.1-2-5　B-scan

（3）C 型显示 C-scan

C 型显示又称 C 扫描或 C 超，它以反射回波作为辉度调制信号，用亮点或暗点显示接收信号，缺陷回波在荧光屏上显示的亮点构成了被检测对象中缺陷的平面投影图，如图 1-2-6 所示。这种显示方式能给出缺陷的水平投影位置，但不能确定缺陷的深度。

C-mode display also known as C-scan or C-ultrasound, uses the reflected echo as a luminance modulation signal and displays the received signal with bright or dark spots. The bright spots of defect echoes on the fluorescent screen form a two-dimensional projection map of the defects in the object being inspected, as shown in Fig.1-2-6. This display method can provide the horizontal projection position of the defect, but cannot determine the depth of the defect.

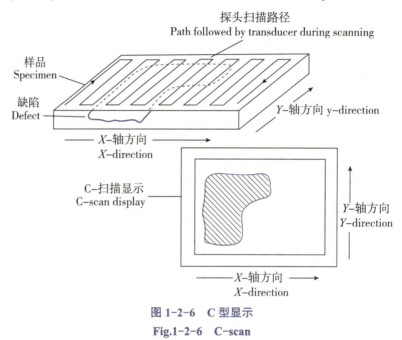

图 1-2-6　C 型显示

Fig.1-2-6　C-scan

思考：图 1-2-7 对应的超声显示形式是什么？

Thinking: Which kind of scan display does the Fig.1-2-7 belong to?

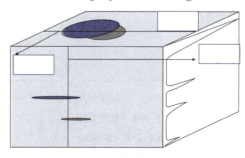

图 1-2-7　超声显示形式

Fig.1-2-7　Ultrasonic Display Form

小任务 4　超声检测的物理原理
Subtask 4　Physical Principles of Ultrasonic Testing

超声检测是基于弹性波在固体材料中传播，并在界面和内部缺陷或损伤（材料分离、夹杂物等）处反射。检测过程：将超声波传输到测试材料中，观察反射波（脉冲回波技术）或透射波（透传技术）。在检测中，脉冲回波系统更有用，因为它只需要单侧访问被测试的元件。

Ultrasonic testing T is based on the fact that elastic waves propagate in solid materials and are reflected at the interfaces and by internal flaws or damage (material separations, inclusions). Detection Process:transfer the ultrasonic waves to the test material and observe the reflect wave (Pulse-Echo technique), or the transmitted waves (Through Transmission technique). In practice, Pulse-Echo systems are more useful since they require only one-sided access to the element being tested.

最常见的超声检测方法应用纵向和横向（剪切）波，然而，也使用其他形式的弹性波传播，包括瑞利波和兰姆波。大多数超声波换能器都是基于压电原理工作的，当它们被极短的放电激发时，会发射超声脉冲；当它们接收到返回的超声信号时，会产生电信号。这种相互转换是通过在相对表面上具有电极的极化陶瓷或晶体材料板来实现的。传感器设计用于接触测试或浸入测试。接触换能器需要在探针和测试元件表面之间施加耦合剂（如水、薄膜或油、接触膏或甘油），因为空气衰减了大部分超声能量。在测试中，弹性波通过水路或液柱传输到材料中。由于声波的差异，可以探伤各种材料的阻抗，这是对超声波在材料中传播的阻力的测量。当声学阻抗发生失配时，超声波在两种不同材料的界面或其他不连续处反射。

The most common ultrasonic testing methods apply longitudinal and transverse (shear) waves, however, other forms of the elastic waves propagation are also used, including Rayleigh and Lamb waves. Most ultrasonic transducers operate on the piezoelectric principle of transconductance, i.e. they transmit an ultrasonic pulse when excited by an extremely short electrical discharge, and generate an electrical signal when they receive the returned ultrasonic signal. This mutual conversion is accomplished by a plate of polarized ceramic or crystalline material with electrodes on the opposite surfaces. Transducers are designed for contact testing or immersion testing. Most contact transducers require applying a coupling agent (e.g. water, thin film or oil, contact paste, or glycerin) between the probe and the test element surface because air attenuates most of the ultrasonic energy. In the immersion testing, the elastic waves are transmitted into the material through a water path or a liquid column. Flaw detection is possible because of the difference in acoustic impedance of various materials, which is a measure of resistance to propagation of ultrasonic wave in a material. Ultrasonic waves are reflected at the interface of two different materials or other discontinuities, when the mismatch in acoustic impedance occurs.

另外，空气耦合超声检测技术是另一种新型超声技术，通过消除液体耦合剂可检测带有穿孔面板的蜂窝夹芯等结构。空气耦合超声的最大挑战是需要使用特殊换能器补偿气固界面处超声能量的大幅损失。

It should be mentioned that there is also an air-coupled ultrasonic testing technique. It eliminates the necessity for a liquid couplant and can be applied to such structures as honeycomb sandwich with perforated face sheets. The greatest challenge in air-coupled ultrasonic testing is the enormous loss of ultrasonic energy at the air-solid interface; therefore, special transducer types are used.

小任务 5　声学表征量的参数
Subtask 5　Parameters for Acoustic Characterization

（1）声速 Sound Velocity

超声波的传播速度简称为声速。声速是一个重要的声学参数，它取决于传声介质自身的密度、弹性模量等性质，还与超声波的波形有关。对于纵波、横波和表面波来说，每种波形的声速值仅与传声材料的性质有关。层合板的声速较为复杂，除与材料特性相关外，还与频率、板厚和振动模式有关。了解材料的声速，对于缺陷的定位、定量分析以及测厚等具有重要意义。

The propagation speed of ultrasonic waves is referred to as sound velocity. Sound velocity is an important acoustic parameter that depends on the properties of the propagation medium, such as density and elastic modulus, and is also related to the waveform of the ultrasonic wave. For longitudinal, transverse, and surface waves, the sound velocity values for each waveform depend only on the properties of the transmission medium. The sound velocity of plate waves is more complex, as it depends not only on the material characteristics but also on the frequency, plate thickness, and vibration mode. Understanding the sound velocity of materials is of great significance for defect localization, quantitative analysis, thickness measurement, etc.

纵波、横波和表面波的声速与介质自身性质之间的关系如下。

The relationships between the sound velocity of longitudinal, transverse, and surface waves and the properties of the medium itself are as follows.

纵波在固体介质中的声速：

The sound velocity of longitudinal waves in solid medium:

$$CL=\sqrt{\frac{E}{\rho}}\sqrt{\frac{1-\sigma}{(1+\sigma)(1-2\sigma)}} \qquad (1)$$

对于钢材，泊松比 $\sigma \approx 0.25$，则 $CL_{钢} = 1.1\sqrt{E/\rho}$；对于铝材，$\sigma \approx 0.33$，则 $CL_{铝} = 1.2\sqrt{E/\rho}$。由此可以看出：声速与弹性模量 E 成正比，与密度 ρ 成反比。

For steel, with a Poisson's ratio $\sigma \approx 0.25$, $CL_{steel} = 1.1\sqrt{E/\rho}$; for aluminum, $\sigma \approx 0.33$, $CL_{Al}= 1.2\sqrt{E/\rho}$. This is because the speed of sound is directly proportional to the elastic modulus E and inversely proportional to the density ρ.

纵波在液体和气体介质中的声速：

The speed of longitudinal waves in liquid and gas media:

$$CL=\sqrt{\frac{B}{\rho}} \qquad (2)$$

其中，B 表示液体和气体介质的体积弹性模量 (体积膨胀系数)。

Where B represents the bulk modulus (volumetric elasticity) of liquid and gas media.

横波在固体介质中的声速：

The velocity of transverse waves in solid media:

$$CS=\sqrt{\frac{G}{\rho}}=\sqrt{\frac{E}{\rho}}\sqrt{\frac{1}{2(1+\sigma)}} \qquad (3)$$

其中，G 表示介质的切变模量 (单位面积上的切向应力与变形角之比)。对于钢材，$CS/CL \approx 0.58$；对于铝材，$CS/CL \approx 0.50$。

Among them, G represents the shear modulus of the medium (the ratio of tangential stress to deformation angle per unit area). For steel, $CS/CL \approx 0.58$; for aluminum, $CS/CL \approx 0.50$.

表面波在固体介质表面传播的声速：

The velocity of surface wave propagation on the surface of a solid medium:

$$CR=\frac{0.87+1.112\sigma}{1+\sigma}\sqrt{\frac{G}{\rho}}=\frac{0.87+1.112\sigma}{1+\sigma}\sqrt{\frac{E}{\rho}}\sqrt{\frac{1}{2(1+\sigma)}} \qquad (4)$$

对于钢材，$CR/CS \approx 0.92$；对于铝材，$CR/CS \approx 0.93$。

For steel, $CR/CS \approx 0.92$; for aluminum, $CR/CS \approx 0.93$.

表 1-2-1 中所列的是几种常用材料的密度、声速和波长值。可以看出，材料的差异引起声速的变化非常明显。对于钢和铝而言，横波声速约为纵波声速的一半，而表面波声速约为横波声速的 90%。另外，纵波、横波和表面波声速不随频率而变化，因此，在给定的材料中，频率越高，波长越短。

The Table 1-2-1 lists the density, acoustic velocity, and wavelength values of several commonly used materials. It can be seen that the differences in materials cause significant changes in acoustic velocity. For steel and aluminum, the transverse wave velocity of sound of sound is about half of the longitudinal wave velocity of sound, while the surface wave velocity of sound is about 90% of the transverse wave velocity of sound. In addition, the longitudinal, transverse, and surface wave velocities do not vary with frequency. Therefore, in a given material, the higher the frequency, the shorter the wavelength.

表 1-2-1　不同材料的密度、声阻抗和 5 MHz 时的波长
Table 1-2-1　Different material density, acoustic impedance and wavelength for 5 MHz

材料 Material	密度 /(g · cm^{-3}) Density/(g · cm^{-3})	声阻抗 /(10^6 kg · m^{-2} · s^{-1}) Acoustic impedance	纵波 L		横波 S	
			CL/(m · s^{-1})	λ/mm	CS/(m · s^{-1})	λ/mm
铝 Al	2.70	17	6 300	1.30	3 100	0.63
钢 Steel	7.80	45	5 900	1.20	3 200	0.64
钛 Ti	4.50	27	6 100	1.22	3 110	0.62
复合材料 Composite	1.54	4.7	3 000	0.60	—	—
有机玻璃 Acrylic glass	1.18	3.2	2 700	0.50	1 120	0.22
水 (20 ℃) Water	1.00	1.5	1 500	0.50	—	—
空气 Air	0.001 2	0.004 3	340	0.07	—	—

　　尽管各向同性均匀介质对应于特定材料、波形、声速值为一个常数。但是，当介质本身存在不均匀性，以及介质发生温度、应力等变化时，介质的密度、弹性性质会有相应的变化，从而会引起声速的改变。较为普遍的是介质温度的改变对声速的影响。通常情况下，随温度升高，都会引起声速降低。例如，图 1-2-8 中左图是有机玻璃和聚乙烯材料中声速与温度的关系。

Although isotropic homogeneous media correspond to a specific material, waveform, and sound speed is a constant, the density and elastic properties of the medium will change in response to variations in temperature, stress, and other factors. This will cause changes in the sound speed. The most common cause of changes in sound speed is variations in temperature. Typically, as temperature increases, sound speed decreases. For example, the left graph in Fig.1-2-8 shows the relationship between sound speed and temperature in materials such as acrylic and polyethylene.

材料的非均匀性引起声速的变化是另一个需要注意的问题。如奥氏体不锈钢粗晶材料的检测，由于声束在穿过不均匀的粗大晶粒的晶界时，声速的变化可使声束方向偏离原方向传播，从而使检测结果受到影响。

The variation of sound velocity due to the non-uniformity of materials is another issue that needs attention. For example, in the inspection of coarse-grained austenitic stainless steel materials, the change in sound velocity caused by the variation of grain size can make the direction of the sound beam deviate from the original direction when passing through the uneven coarse grains and grain boundaries, thus affecting the inspection results.

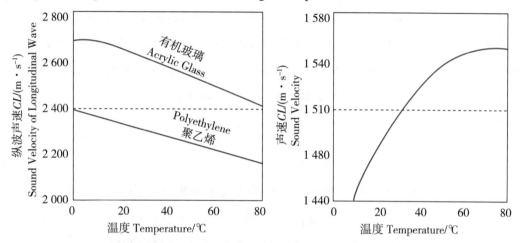

图 1-2-8　有机玻璃、聚乙烯中声速与温度关系（左）和水中声速与温度关系（右）

Fig.1-2-8　**Relationship between Sound Velocity and Temperature in Acrylic Glass, Polyethylene (Left), and Water (Right)**

另外，材料组织结构和排列差异也是引起声速变化的因素。例如，对于碳纤维环氧复合材料而言，声波沿垂直于纤维方向传播的声速约为 3 000 m/s，而沿纤维方向传播的声速为 8 400 m/s，不同方向的声速相差很大。材料的非均匀性或组织差异引起的声速变化，可以使声速测量技术作为材料评价的一种手段。例如，通过声速测定可以确定复合材料的纤维取向，检验复合材料强度性能、玻璃纤维复合板中纤维的质量分数，以及粉末冶金材料的致密性等。

In addition, differences in material organization and arrangement are also factors that cause changes in sound velocity. For example, for carbon fiber reinforced epoxy composite materials, the sound velocity perpendicular to the fiber direction is about 3 000 m/s, while the sound velocity along the fiber direction is 8 400 m/s, and the sound velocity in different directions differs greatly. The variation of sound velocity caused by the non-uniformity or organization difference of materials also makes sound velocity measurement technology a means of material evaluation. For example, through sound velocity measurement, the fiber orientation of composite materials can be determined, the strength and performance of composite materials can be tested, the fiber concentration in glass fiber composite panels can be determined, and the density of powder metallurgical materials can be evaluated.

（2）声压与声阻抗 Sound Pressure and Acoustic Impedance

声压是指在声波传播的介质中，某一点在某一时刻所具有的压强与没有声波存在时该点的静压强之差。声压单位是 Pa（帕斯卡），用 p 表示。声场中，每一点的声压是一个随时间和距离变化的量，可用下式表达：

Sound pressure refers to the difference between the pressure at a certain point in a medium during the propagation of a sound wave and the static pressure at that point in the absence of a sound wave. The unit of sound pressure is Pascal (Pa) and is represented by p. In a sound field, the sound pressure at each point is a quantity that varies with time and distance, and can be expressed using the following formula:

$$p = \rho c A \omega \sin \omega \left(t - \frac{x}{c} \right) = \rho c u \tag{5}$$

其中，ρ 表示介质的密度；c 表示介质的声速；A 表示质点的位移振幅；ω 表示角频率；u 表示质点振动速度。式中，$\rho c A \omega$ 或 $\rho c u$ 是声压的振幅。在实际应用中，A 扫描超声检测仪中屏幕显示信号如图 1-2-9 所示，比较两个声波并不需要对每时刻 t 的声压进行比较，真正代表超声波强弱的是声压幅度。因此，通常将声压幅度简称为声压，也用符号 p 表示，则 $p=\rho c A \omega$。超声检测仪荧光屏上脉冲信号幅度与声压成正比，因此，通常读出的信号幅度的比等于声压比。

In the equation, ρ represents the density of the medium; c represents the speed of sound in the medium; A represents the displacement amplitude of the particle; ω represents the angular frequency; u represents the particle's oscillation velocity. In the equation, $\rho c A \omega$ or $\rho c u$ is the amplitude of the sound pressure. In practice, In practice, signals shown in forms of A-scan is shown as Fig.1-2-9, comparing two sound waves does not require comparing the sound pressure at every moment t. What truly represents the strength of an ultrasound wave is the sound pressure amplitude. Therefore, sound pressure amplitude is often referred to as sound pressure, which is also represented by the symbol p. Thus, $p=\rho c A \omega$. The amplitude of the pulse signal on the fluorescent screen of an ultrasonic instrument is proportional to the sound pressure. Therefore, the ratio of the signal amplitudes read out is equal to the ratio of the sound pressures.

另外，由 $p=\rho c u$ 可知，在同一声压的情况下，ρc 越大，质点振动速度 u 越小；反之，质点振动速度 u 越大，所以把 ρc 称为介质的声阻抗，用 Z 表示。声阻抗的单位是 kg/(m² · s)。声阻抗能直接表示介质的声学性质，在超声检测领域所采用的许多方程式中经常出现的是介质密度与声速的乘积而不是其中某一个值，因此，常将 ρc 作为一个独立的概念来理解。声阻抗决定着超声波在通过不同介质的界面时能量的分配，因此，它是一个重要的声学参数。

In addition, as $P=\rho cu$, it can be seen that for the same sound pressure, the larger the product of ρc, the smaller the particle vibration speed u, and vice versa. Therefore, ρc is called the acoustic impedance of the medium, and is represented by the symbol Z. The unit of acoustic impedance is kg/(m² · s). Acoustic impedance can directly represent the acoustic properties of the medium. In many equations used in ultrasonic testing, the product of medium density and sound velocity is often used instead of just one of these values. Therefore, ρc is often understood as an independent concept. Acoustic impedance determines the distribution of energy when ultrasonic waves pass through interfaces of different media, making it an important acoustic parameter.

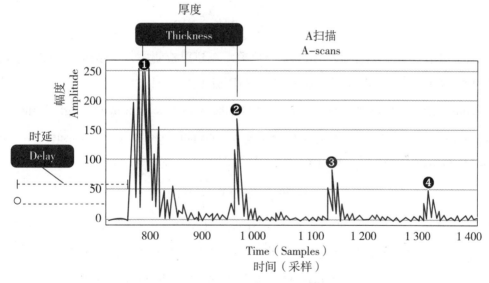

图 1-2-9　A 扫描超声检测仪中屏幕显示信号

Fig.1-2-9　Signals shown in forms of A-scan

（3）声强与分贝 Sound Intensity and Decibels

声强是指在垂直于声波传播方向的平面上，单位时间内通过单位面积的声能平均值，因此，也称为声的能流密度。对于谐振波，常将一周期中能流密度的平均值作为声强，声强的单位为 W/cm²，用符号 I 表示。

Sound intensity refers to the average value of sound energy passing through a unit area per unit time in a plane perpendicular to the direction of sound propagation. Therefore, it is also called sound energy density. For harmonic waves, the average value of energy density over one period is often used as sound intensity, and the unit of sound intensity is W/cm^2, denoted by the symbol I.

$$I = \frac{P^2}{2\rho c} \tag{6}$$

分贝是用来量度声音的一种声学单位，用符号 dB 表示。它是由比它大 10 倍的"贝耳"导出的。由于人耳对声音响度的感觉近似地与声强的对数成正比，于是采用某一声音强度 I 与基准声强（人耳感觉的最小声强 I_0）之比的对数来表示这一关系，即

Decibel (dB) is a unit of measurement used to quantify sound in acoustics. It is represented by the symbol dB and is derived from the 'bel', which is 10 times larger than it. Since the human perception of loudness is approximately proportional to the logarithm of sound intensity, a logarithmic scale is used to express the relationship between a certain sound intensity I and a reference sound intensity (the minimum sound intensity perceived by the human ear, denoted as I_0). The formula is:

$$\varDelta = 10 \lg \frac{I}{I_0} = 20 \lg \frac{P}{P_0} (\text{dB}) \tag{7}$$

式中，$\varDelta$ 表示 I 与 I_0 或者 P 与 P_0 的分贝差。在超声检测中，当超声检测仪的垂直线性较好时，显示屏上的信号幅度 (波高)H 与声压成正比，所以

In ultrasonic testing, when the vertical linearity of the ultrasonic instrument is good, the signal amplitude (wave height) H displayed on the screen is proportional to sound pressure. Therefore, the sound intensity can be expressed as:

$$\varDelta = 20 \lg \frac{P_2}{P_1} = 20 \lg \frac{H_2}{H_1} (\text{dB}) \tag{8}$$

其中，声压基准 P_1 或幅度基准 H_1 可以任意选取。当幅度比 H_2/H_1=1 时，$\varDelta$=0 dB，表明两个信号幅度相等时，分贝差为 0 ；当 H_2/H_1=2 时，$\varDelta$=6 dB，称幅度 H_2 比幅度 H_1 高 6 dB ；当 H_2/H_1=1/2 时，$\varDelta$= −6 dB，称幅度 H_2 比幅度 H_1 低 6 dB。 在超声检测时，分贝值用处非常广泛，如调整检测灵敏度时，可用分贝值表示可检测信号幅度与参考试块中人工缺陷反射幅度的关系；进行缺陷评定时，可用分贝值将缺陷显示幅度与人工缺陷反射幅度进行比较，表示缺陷显示幅度的大小。

The reference sound pressure P_1 or amplitude reference H_1 can be arbitrarily selected. When the amplitude ratio H_2/H_1 is 1, $\varDelta$=0 dB, indicating that the decibel difference is 0 when the two signal amplitudes are equal. When H_2/H_1=2, $\varDelta$=6 dB, which means that the amplitude H_2 is 6 dB higher than H_1. When H_2/H_1=1/2, $\varDelta$= −6 dB, which means that the amplitude H_2 is 6 dB lower than H_1. In ultrasonic testing, the decibel value has a wide range of applications. For example, when adjusting detection sensitivity, decibel values can be used to express the relationship between the detectable signal amplitude and the amplitude of the artificial defect reflection in the reference block. During defect assessment, decibel values can be employed to compare the defect display amplitude with the amplitude of the artificial defect reflection, indicating the magnitude of the defect display amplitude.

工单 Work Order

耗材 / 工具 / 设备清单
Material/Tool/Equipment List

耗材 / 工具 / 设备 Material/Tool/Equipment	图示 Illustration
复合材料检测板 Composite Material Inspected Panel 探头 Probe	
检测工具：Detection Tool: 超声探伤仪 Ultrasonic Flaw Detector	

复合材料层合板超声 A 扫描探伤校准程序任务工单

Ultrasonic Inspection A-Scan Calibration Procedure for CFRP Task Card

任务名称 Task Topic 复合材料层合板超声检测 Ultrasonic Inspection For CFRP	
步骤 Steps	规章 / 指令 Regulation/Instruction
1. 适用性说明 Applicability	CFRP 层合板和蜂窝夹芯结构的超声 A 扫描探伤，适用于蒙皮厚度为 0.25 in 以内且只有一个面可达的情况，用于检测结构内部分层。 To detect internals, delamination in graphite/epoxy solid laminate structure and skin over core up to 0.25-inch thick, where only one surface of the part is accessible for inspection.
2. 设备耗材准备 Equipment and Consumable Material Preparation	可以使用任何具有显示信号半波形能力的仪器，以及满足本程序中规定的信号分辨率要求的任何高阻尼探头。 Any instrument with the ability to display half waveforms of a signal, and any highly damped transducer that satisfies the signal resolution requirement. 探头——频率 10 MHz, 0.25 in 直径，有机玻璃延迟块。 Transducer—10 MHz, 0.25-inch diameter, lucite delay line. 耦合剂——任何与 CFRP 结构兼容的耦合剂。 Couplant—Any couplant compatible with graphite/epoxy structure.
3. 仪器分辨率检查 Instrument Resolution Check	在校准试块表面涂上一层耦合剂。 Apply a thin film of couplant to surface of calibration guides. 根据操作手册进行初步的仪器调整。 Perform preliminary instrument adjustments per owner's operating manual. 如果可能，同时显示负半波形和正半波形。使用单尖峰波形。 If possible, display both the negative half waveform and the positive half waveform. Use waveform with single spike. 将探头放置在校准试块的第 30 层位置。使用延迟块和范围控制将前表面反射位置置于屏幕左侧边缘，将后表面反射位置置于屏幕宽度的 90%，如图 1-2-10 所示。 Place transducer on step 30 of Calibration Guide. Use the delay and range controls to position the front surface reflection at the left edge of screen and the back surface reflection at 90 percent screen width. (Fig. 1-2-10). 图 1-2-10 　30 层的校准信号 Fig.1-2-10 　Calibration signal at layer 30 注意：屏幕范围现在设置为检查最大 0.25 in 厚的碳纤维 / 环氧树脂层合板。 Note: Screen range is now set to inspect a maximum of 0.25-inch thick graphite/epoxy laminate.

任务名称 Task Topic 复合材料层合板超声检测 Ultrasonic Inspection For CFRP	
步骤 Steps	规章 / 指令 Regulation/Instruction
	将探头置于校准试块的第 15 层上。调整增益，以使背面回波达到大约 50% 的屏幕高度，如图 1-2-11 所示。 Place transducer on step 15 of Calibration Guide. Adjust gain to obtain a back surface reflection of approximately 50 percent of full screen height (Fig. 1-2-11). 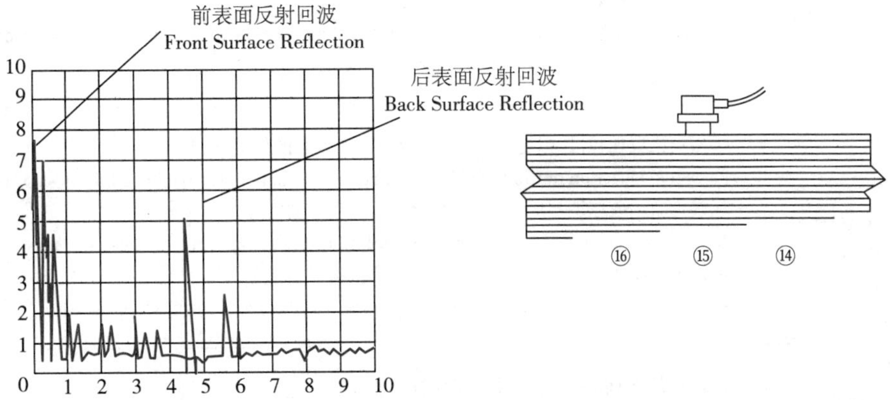 图 1-2-11　调整 15 层的回波信号 Fig.1-2-11　Adjustment of the echo signal at layer 15
3. 仪器分辨率检查 Instrument Resolution Check	将探头置于校准指南的第 1 层上。仪器 / 探头的分辨率必须能够很容易识别出背面回波，如图 1-2-12 所示。 Place transducer on step 1 of Calibration Guide. Instrument/ transducer resolution must be such that an easily definable back surface reflection is obtained. 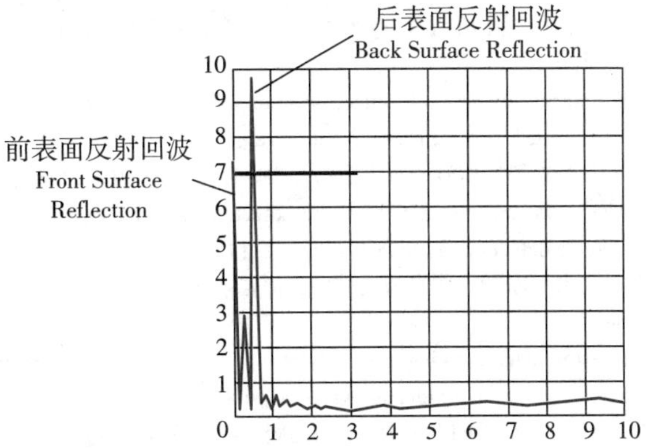 图 1-2-12　调整 1 层的回波信号 Fig.1-2-12　Adjustment of the echo signal at layer 1 将探头沿校准试块从 30 层开始到 1 层进行扫查。记录在任意一层发生厚度信号移动和厚度分辨率变化的情况。 Position the transducer along the calibration guides from step 30, through all steps to step 1. Note the signal shift and ply resolution at all ply thicknesses.

复合材料层合板超声检测任务工单
Ultrasonic Inspection for CFRP Task Card

任务名称 Task Topic	
复合材料层合板超声检测 Ultrasonic Inspection for CFRP	
步骤 Steps	规章 / 指令 Regulation/Instruction
1. 检查准备 Preparation for Inspection	清除松动的漆层，并彻底清洁检查表面。 Remove loose paint and thoroughly clean inspection surfaces. 确保表面粗糙度满足探伤无障碍的要求。如有必要，可用特氟龙擦洗垫（即 Scotchbrite 或同等材料）轻轻擦拭表面。 Ensure inspection surface is free from obstructing roughness. If necessary, abrade surface lightly with a teflon scrubbing pad, i.e., Scotchbrite or equivalent.
2. 仪器校准 Instrument Calibration	将探头放置在校准试块上最接近层合板待检区域的位置。 Place transducer on portion of calibration guide closest to noted laminate thickness of inspection area. 观察背面回波在屏幕时基线的水平位置。 Note horizontal position of the back surface reflection along scope baseline. 调整增益以获得 100% 全屏高度的背面回波。 Adjust gain to obtain a back surface reflection of 100 percent of full screen height.
3. 检测程序 Inspection Procedure	扫描所有位于同一个层合板厚度的检验区域。 Scan all inspection areas of the same laminate thickness. 注：背面回波信号振幅可能在屏幕高度的 20%~100% 变化。 Note: Back surface signal amplitude may vary between 20 percent and 100 percent of screen height.
4. 检验结果 Inspection Results	由各种结构或材料条件引起的信号变化可能被错误地解释为缺陷状况。通常是三种状态。 Signals caused from various construction or material conditions can be erroneously interpreted as defect conditions. （1）背面回波在预期的屏幕位置完全消失，且在预期的背面回波位置右侧出现一个新信号，如图 1-2-13 所示。 100 percent loss of back reflection at expected screen position with appearance of a new signal to the right of expected back reflection position (Fig.1-2-13). 预期信号右侧的信号响应表明层合板厚度增加，不被认为是缺陷条件，如图 1-2-14 所示。 A signal response to the right of the expected signal indicates an increase in laminate thickness and is not considered a defect condition, see Fig. 1-2-14. 此时需参考图纸，确定铺层结构。铺层搭接定义为增加一层厚度，其宽度不超过 1 in，并呈直线形延伸。 Check reference drawings for a ply buildup. Ply overlaps can be defined by a one ply thickness increase not exceeding 1 in. in width and extending in a straight line. 先前修复的区域可以通过蒙皮表面圆形或矩形补丁的目视检查来确定，或者通过将探头从良好区域移动到可疑区域，并观察圆形或矩形补丁边缘的单层变化来确定。 Previously repaired areas can be defined by visual evidence of a round or rectangular patch on skin surface or by moving probe from good area into suspect area and observing a one ply change around the edge of a round or rectangular patch. （2）信号响应显示层合板厚度减小，而参考图纸上未予注明，表明存在缺陷。 Signal response indicating a decrease in laminate thickness not noted in reference drawings is a defect condition.

任务名称 Task Topic 复合材料层合板超声检测 Ultrasonic Inspection for CFRP	
步骤 Steps	规章 / 指令 Regulation/Instruction
4. 检验结果 Inspection Results	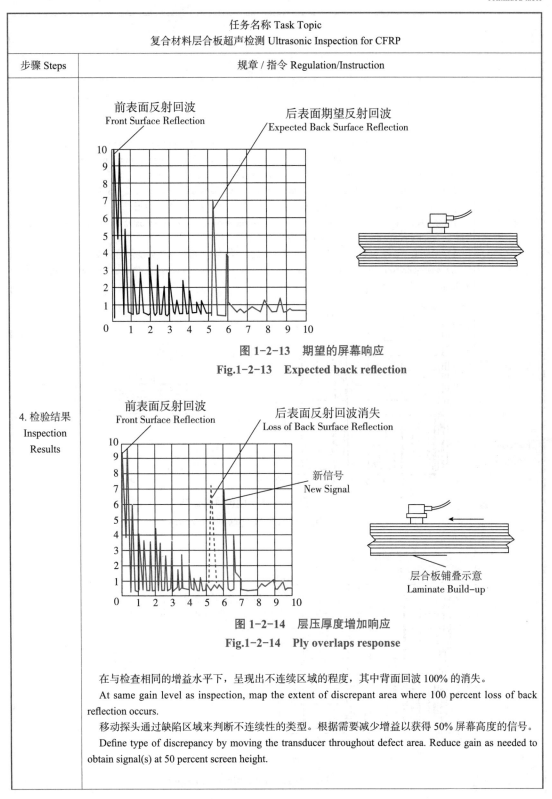 前表面反射回波 Front Surface Reflection　　后表面期望反射回波 Expected Back Surface Reflection 图 1-2-13　期望的屏幕响应 Fig.1-2-13　Expected back reflection 前表面反射回波 Front Surface Reflection　　后表面反射回波消失 Loss of Back Surface Reflection 新信号 New Signal 层合板铺叠示意 Laminate Build-up 图 1-2-14　层压厚度增加响应 Fig.1-2-14　Ply overlaps response 在与检查相同的增益水平下，呈现出不连续区域的程度，其中背面回波 100% 的消失。 At same gain level as inspection, map the extent of discrepant area where 100 percent loss of back reflection occurs. 移动探头通过缺陷区域来判断不连续性的类型。根据需要减少增益以获得 50% 屏幕高度的信号。 Define type of discrepancy by moving the transducer throughout defect area. Reduce gain as needed to obtain signal(s) at 50 percent screen height.

任务名称 Task Topic 复合材料层合板超声检测 Ultrasonic Inspection for CFRP	
步骤 Steps	规章 / 指令 Regulation/Instruction
4. 检验结果 Inspection Results	在屏幕上预期信号左侧有某单个信号，表示存在分层缺陷。对比校准试块响应信号在沿屏幕时基线同一水平位置，来计算缺陷深度。 A single signal to the left of the expected signal at CRT screen indicates a delamination. Note depth by comparison with calibration guide response obtained at the same horizontal position along scope baseline. 屏幕上出现多个信号表明层合板断裂，如图 1-2-15~ 图 1-2-17 所示。 Multiple signals on CRT screen indicate a fracture in the laminate, i.e., Fig. 1-2-15-Fig.1-2-17. （3）如果背面回波低于屏幕高度的 20%，则表明存在潜在的缺陷，可通过以下情况确认： If the back surface reflection is less than 20% of the screen height, it indicates a potential defect, which can be confirmed by the following conditions: 1）增加增益以获得 50% 屏幕高度的背面回波，观察是否可以得到分辨背面回波。 Increase gain to obtain a back surface reflection of 50 percent screen height if a definable back surface reflection is obtained. 图 1-2-15　期望屏幕响应 Fig.1-2-15　Expected Response 图 1-2-16　分层响应 Fig.1-2-16　Delamination response

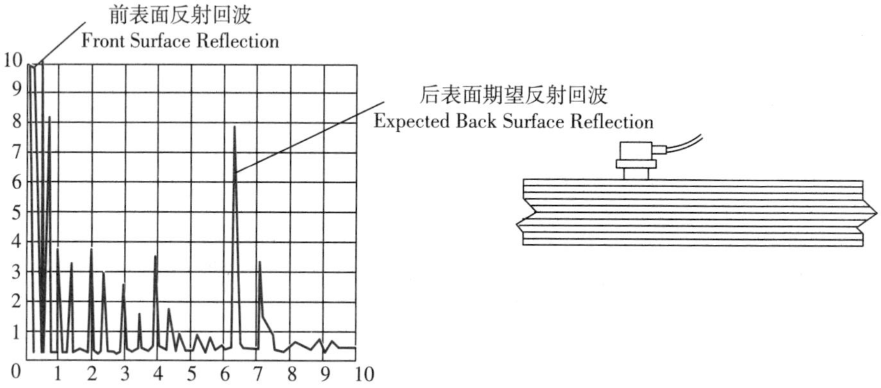

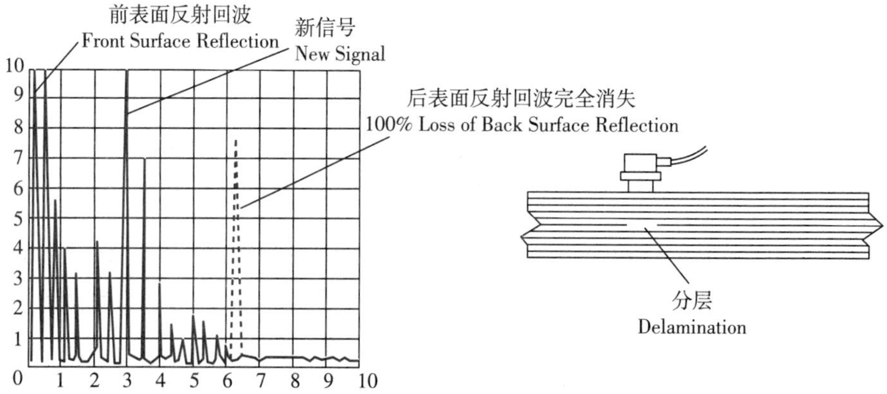

任务名称 Task Topic 复合材料层合板超声检测 Ultrasonic Inspection for CFRP	
步骤 Steps	规章 / 指令 Regulation/Instruction
4. 检验结果 Inspection Results	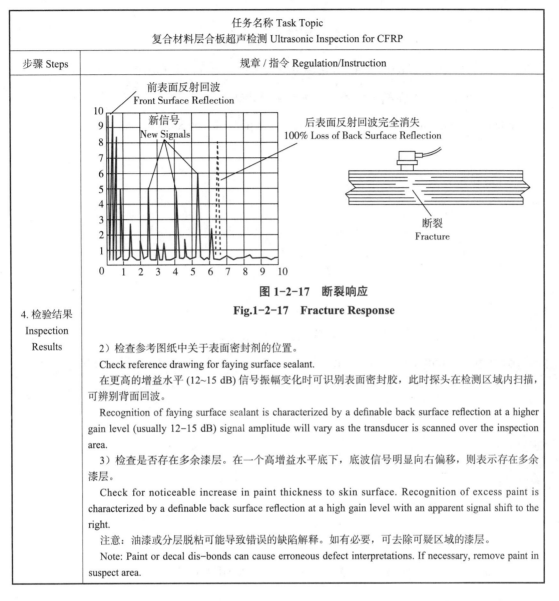 图 1-2-17　断裂响应 **Fig.1-2-17　Fracture Response** 2）检查参考图纸中关于表面密封剂的位置。 Check reference drawing for faying surface sealant. 在更高的增益水平 (12~15 dB) 信号振幅变化时可识别表面密封胶，此时探头在检测区域内扫描，可辨别背面回波。 Recognition of faying surface sealant is characterized by a definable back surface reflection at a higher gain level (usually 12−15 dB) signal amplitude will vary as the transducer is scanned over the inspection area. 3）检查是否存在多余漆层。在一个高增益水平底下，底波信号明显向右偏移，则表示存在多余漆层。 Check for noticeable increase in paint thickness to skin surface. Recognition of excess paint is characterized by a definable back surface reflection at a high gain level with an apparent signal shift to the right. 注意：油漆或分层脱粘可能导致错误的缺陷解释。如有必要，可去除可疑区域的漆层。 Note: Paint or decal dis−bonds can cause erroneous defect interpretations. If necessary, remove paint in suspect area.

超声检测练习题任务工卡

Ultrasonic Inspection Exercise Task Card

（1）什么是声程？在超声检测中，如何测量声程？

What is sound path? How to measure the sound path in ultrasonic testing?

（2）在进行超声检测时，如何选择合适的探头和频率？

When performing ultrasonic testing, how to choose the appropriate probe and frequency?

（3）简要说明 A 扫描和 B 扫描在超声检测中的应用和原理。

Briefly explain the applications and principles of A-scan and B-scan in ultrasonic testing.

子任务 3 记录损伤信息
Subtask 3 Damage Information Report

【子任务解析 Subtask Analysis 】

航线结构损伤快速处理方法指出：目视检查与敲击检测所有记录的分层／脱粘区域，必须按照 SRM 要求进行验收或修复行动的评估，并形成规范结构损伤报告，MCC 技术支援工程师或结构工程师对此报告进行评估并处理。

In accordance with SRM requirements of rapid airline structural damage treatment, visual check and tapping test detection results for any delamination areas must be evaluated for acceptance or repairs operation. A standardized structural damage report is also required. MCC technical support engineer or structural engineer will make evaluation and treatment refer to the damage report.

通过本任务的训练，可掌握规范书写结构损伤报告的方法，理解可允许损伤的标准，进而确定最小修理范围与修理方案。

Through the training of this task, students can master the method of writing structural damage report. Acknowledge the standard of allowable damage, and then determine the minimum repair range and repair plan.

【子任务分解 Subtask Break-down 】

（1）确定损伤尺寸的方法。

Method of determining damage size.

（2）结构损伤报告规范书写。

Standard writing of structural damage report.

小任务 1　航线损伤报告主要包含哪些信息？
Subtask 1　What information does the damage report contain?

损伤报告的内容应包括但不限于以下信息。

The content of the damage report shall include but not limited to the following information.

（1）飞机基本信息和当前状况。

Basic information and status of the aircraft.

（2）损伤部件、损伤类型与损伤成因。

Damaged parts, types and causes of damage.

（3）损伤站位信息。

Damaged station information.

（4）损伤尺寸、边距、间距信息，如距离最近紧固件、桁条、隔框的距离，损伤之间的距离等。

Damage size and distance information, such as the distance from the nearest fastener, stringer, frame, distance between damages, etc.

（5）执行了何种检查方式及检查结果。

Inspection procedure has been performed and the inspection results.

（6）其他信息（特别是影响可允许损伤判断的关键数据）。

Other information (especially the key data affecting the determination of allowable damage limit).

小任务 2　如何理解站位信息？
Subtask 2　How to understand the station information?

首先，飞机的站位与平板件的站位是不同的。飞机的站位信息指飞机的坐标，通过在空间指定 XYZ 轴以定位飞机构件。飞机站位源于工位，其定义是基于机体装配提出的。由于飞机的机身、机翼、尾翼等大型构件通常是异地生产，运送后再进行装配，因此需要精确定位来"解剖"飞机。STA 水平站位、WL 水线、BL 纵剖线即代表 X、Z、Y。因此，飞机任意位置出现损伤，都可以通过查询站位信息进行精确定位。

First, the dimension information or station information of the aircraft is different from that of the flat parts. Station of the aircraft, referring to the coordinates of the aircraft, by specifying the XYZ axis in space to locate the aircraft components. The aircraft station is derived from the station and its definition is put forward based on the assembly of the airframe. Because large components such as the fuselage, wings and tail of an aircraft are often manufactured in different locations, shipped and assembled, precise positioning is needed to "dissect" the aircraft. The STA horizontal station, WL waterline and BL vertical section line represent X, Z and Y. Therefore, damage at any position of the aircraft can be accurately located by querying station information.

针对二维构件，其站位信息已由三维简化为二维，因此针对存在损伤的平板件，损伤的站位信息可归纳为三种表达方式：$(X、Y、Z_0)$，$(X、Y_0、Z)$ 和 $(X_0、Y、Z)$，其中 X_0、Y_0、Z_0 为常数，即指平板件上存在的所有损伤，在某一方向上的站位保持一致。以红豆吐司面包为例，现需要将面包切片，且要求切片包含红豆。二维构件的站位信息可由此类推理解。

As for two-dimensional components, the station information has been simplified from three-dimensional to two-dimensional. Therefore, for plate components with damage, the station information of damage can be summarized into three expressions: (X, Y, Z_0), (X, Y_0, Z) and (X_0, Y, Z), where X_0, Y_0, Z_0 are constants, that is, all the damage on the plate parts, and the station position in a certain direction remains consistent. Take red bean toast as an example. Now you need to cut the bread into slices and require that the slices contain red beans. The station information of two-dimensional components can be understood by analogy.

结构损伤报告任务工卡
Structural Damage Report Task Card

损伤 类型 Damage Type	裂纹 Crack □	凹坑 Dent □	刺穿 Puncture □	分层 / 脱粘 / 空谷 Delamination/ Debond/Void □		腐蚀 Corrosion □
	断裂 Rupture □	屈曲 Buckle □	划伤、缺口、凿伤 Scratch/Nick/Gouge □	其他 Other _____ □		

准确的损伤位置 Exact Damage Position		损伤尺寸 Damage Dimension		主要损伤零件 Main Damaged Part	
站位 Station	mm	长度 Length	mm	件号 Part #	
水线 Water Line	mm	宽度 Width	mm	序列号 Serial #	
纵剖线 Buttock Line	LH□ RH□	深度 Depth	mm	距离上一次大修的时间 Time from the last overhaul	hours

损伤描述 Damage Description	
成因分析 Mechanism	

草图 Damage Sketch / Drawing

（包含参考点坐标、位置、测量尺寸及相邻结构件信息）
(including reference points, location, measurements and adjacent structure as applicable)

损伤评估与记录任务工单
Damage Assessment and Record Task Card

任务名称 Task Topic 损伤评估与记录 Damage Assessment and Record	

步骤 Steps	规章／指令 Regulation/Instruction
1. 操作前清点检查 Tool and material inventory	工具检查：示例 Tool inventory：Example 工具清单完整，定检标识完整，均在有效期内。 Complete tool list. Complete inspection mark. All within the validity period. 材料检查：示例 Material inventory：Example 材料清单完整，合格证明完整。发现固化剂超期，其余均在保质期内。 Complete materials list. Complete certificate of qualification.Curing agent has expired, the rest are within the expiration date.
2. 损伤区域的确定 Determination of structural damage area	(1) 根据 AC-43.13-1B，5 章，5-15~5-24 页，目视检查发现凹坑损伤，直径（　）。 AC-43.13-1B, chapter 5，page 15-24, visual check inspected dent, diameter（　）. (2) 根据 AC-43.13-1B，5 章，8 部分，5~105 页，敲击检测发现存在分层损伤，分层范围约为（　）。 AC-43.13-1B, chapter 5, Section 8，page 5-105, tap testing inspect delamination damage, the delamination range is about（　）.
3. 修理计划的确定 Determination of repair plan	缺陷经打磨确认损伤层数为（　　），根据单层标准厚度理论值（　　），板厚公差为（　　），则损伤范围深度值为（　　）（精确到 2 位有效数字）。 The number of defect layers confirmed by polishing the flaw is（　）. According to the single-layer standard thickness theoretical value（　）and the plate thickness tolerance（　）, the depth value of the damage range is（　）. (Rounded to 2 significant figures). 注意：对于划伤、凿伤、凹坑和蜂窝压陷等类型损伤，需进行规范打磨确认损伤深度，或根据超声检测结果，损伤深度为（　　），宽度为（　　）。 Notes: As for scratches, gouges, dents, honeycomb depression and other types of damage, standard grinding may be required to confirm the depth of damage. Alternatively, based on the ultrasonic inspection results, the damage depth is（　）and the width is（　）.

损伤评估与记录练习题任务工卡
Damage Assessment and Record Exercise Task Card

（1）复合材料中的裂纹如何形成？如何评估和记录复合材料中的裂纹？

How do cracks form in composite materials? How can cracks in composite materials be assessed and recorded?

（2）什么是复合材料的"重要损伤"？如何识别和记录它们？

What is "significant damage" to composite materials? How can they be identified and recorded?

（3）复合材料损伤记录中的信息应包括哪些内容？

What information should be included in the record of composite material damage?

任务 1　损伤评估与记录考核表

姓名		班级		
评价维度	分值 / 分	自评（30%）	互评（30%）	师评（40%）
素养（20%）				
1. 材料清点齐全	5			
2. 任务书自学情况	5			
3. 安全文明操作及 6S	10			
技能（60%）				
1. 敲击检测操作规范性	20			
2. 超声检测操作规范性	20			
3. 损伤报告填写规范性	20			
总结报告（20%）				
1. 训练总结的完整性	5			
2. 训练总结的规范性	5			
3. 个人反思与拟订后续学习计划	10			
总计	100			
任务完成情况	提前完成			
	准时完成			
	滞后完成			

维修方案制定
Maintenance Scheme Determination

【任务情境 Task Scenario 】

您是 WS 航空公司的航线大修结构工程师，收到航线报机翼损伤故障，机身存在两处凹痕。这两个凹痕位于机身右侧，在 FR 24~25，右翼 12~13，现已完成损伤评估，你的工作是制定维修方案，确定修理程序。

You are a structural engineer in charge of major repairs for WS Airlines. You have received a report of wing damage and two dents on the aircraft body. These two dents are located on the right side of the aircraft, between FR 24–25 and between the right wing 12–13. The damage assessment has been completed, and your job is to develop a repair plan and determine the repair procedures.

【任务解析 Task Analysis 】

制定维修方案是进行复合材料结构损伤维修工作的第二步，主要包含 SRM 查询与损伤定级。完成本任务的训练后，学生应实现以下目标：

Developing a repair plan is the second step in conducting composite structure damage repair work, which mainly involves SRM consultation and damage grading. After completing the training for this task, students should achieve the following goals:

知识目标 Knowledge Objectives

（1）理解可允许损伤限制（ADL）的概念。

Understand the concept of Allowable Damage Limit (ADL).

（2）知道 SRM（结构维修手册）查询方法。

Know how to search the SRM structural repair manual.

（3）知道 FRP 损伤等级划分规则。

Understand the classification rules for FRP damage levels.

 能力目标 **Ability Objectives**

（1）能够根据损伤的特点和程度，制定合理的维修方案。

Able to develop reasonable repair plans based on the characteristics and severity of the damage.

（2）熟悉维修规范和标准，熟悉维修记录的编写和管理。

Familiar with maintenance specifications and standards, as well as the preparation and management of maintenance records.

素质目标 **Emotion Objectives**

（1）具备安全意识和质量意识，确保维修工作的安全和质量。

Possess safety and quality awareness, ensuring the safety and quality of maintenance work.

（2）具备团队协作和沟通能力，能够与其他相关部门和人员协调配合，确保维修任务的顺利完成。

Able to work collaboratively and communicate effectively with other departments and personnel, ensuring the successful completion of maintenance tasks.

【任务分解 Task Break-down 】

子任务 1　SRM 查询
Subtask 1　SRM Consultation

【子任务解析 Subtask Analysis】

在采取相应维修措施前，都应仔细查询手册。飞机结构维修手册（SRM）是飞机制造厂家制定的，通常经航空器型号设计批准所在国的适航当局批准。SRM 作为维修单位对飞机结构进行维护和修理的法定技术文件之一，是制定飞机结构维护和修理方案的主要依据。

It is vital to consult the manual carefully before taking corresponding maintenance measures. Structural Repair Manual (SRM) is announced by the aircraft manufacturer and usually approved by the airworthiness authority of the country where the aircraft type design is approved. SRM as one of the legal technical documents for aircraft structure maintenance and repair, is the main basis for maintenance and repair programs arrangement.

成功完成 SRM 查询训练的学生，将具备使用结构维修手册（SRM）的基本技巧，以及具备执行正确的损伤评估、制作正确的损伤报告的能力。

The participant who successfully complete this practical SRM training program will have the basic skills and confidence in how to use the Structural Repair Manual (SRM), how to perform a correct damage assessment, how to make a correct damage report and will be able to select the corrective action and the correct repair procedure.

【子任务分解 Subtask Break-down】

（1）基本掌握 SRM 内容与大纲。

Master the content and outline of SRM.

（2）掌握查阅通用修理程序的方法。

Master the method of searching general repair procedures.

（3）初步理解复合材料结构 ADL 的定级原理。

Preliminarily understand the composite structure ADL rating rules.

小任务 1　SRM 的主要内容有哪些？
Subtask 1　What is the main content of SRM?

SRM 是民航飞机的一种非客户化的手册，其内容涉及飞机结构的识别、允许损伤及修理的信息。每个主要的飞机型号有其单独的 SRM，如 A319、A320、A321 都有各自的 SRM。其中从 SRM 中可查询到以下重要信息：

SRM is a non-customized manual for civil aviation aircraft, which covers the identification of aircraft structure, allowable damage and repair information. Each major aircraft model has its own SRM, for example the A319, A320, and A321 have their own SRM. The following important information can be obtained from the SRM:

（1）结构允许损伤的标准。对于结构不同部位所出现的各类损伤，SRM 做了图解介绍，并给出界定允许的损伤形式和界限值及处置要求。

Standard of allowable structural damage. The SRM provides a graphic description of various types of damage in different parts of the structure, and defines allowable damage forms and limits as well as disposal requirements.

（2）飞机结构修理通用的施工工艺技术。

General procedure for aircraft structure repair.

（3）典型结构或结构件的修理方案。

Repair schemes for typical structures or structural parts.

（4）飞机结构及主要结构元件的图解说明。

Schematic description of aircraft structure and main structural elements.

小任务 2　如何描述三种损伤定级评估方式？针对出现的结构损伤，确定修理方案的基本流程是什么？

Subtask 2　How to describe the three damage classification assessment methods? What is the basic process for determining a repair plan for structural damage?

（1）允许损伤（分为 A、B、C 三类）。

Allowable damage (divided into three categories: A, B and C).

① 临时可允许的损伤。

Temporary allowable damage.

② 带限制的临时可允许损伤。

Temporary allowable damage with limitation.

③ 永久可允许损伤。

Permanently allowable damage.

（2）不允许的损伤（需要采取修理措施）。

Disallowed damage (repair measures need to be taken).

针对机身出现的结构损伤确定修理方案的基本流程包括以下几个步骤，如图 2-1-1 所示。

The basic process of determining the repair plan for structural damage includes the following steps, as shown in Fig.2-1-1.

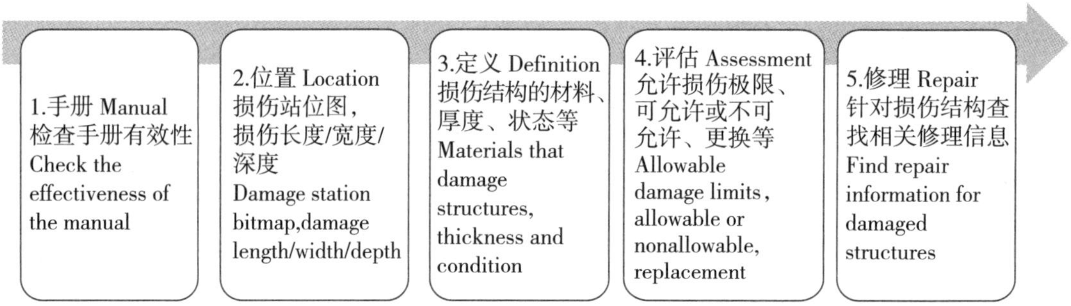

图 2-1-1　修理方案确定通用程序流程图

Fig.2-1-1　General procedure flowchart for determining a repair plan

① 有效性检查：对飞机维修手册、机组手册、飞行手册、训练手册等文献资料进行定期检查和验证，以确保其中的信息和规定符合最新的适航标准和要求，而且能够支持飞机的安全运行。这些手册和资料是飞机运行和维护的重要依据，如果其中的信息不准确或已经过时，可能会对飞机的安全性产生影响。

Conformality check: Refers to the regular inspection and verification of documents such as aircraft maintenance manuals, crew manuals, flight manuals, and training manuals to ensure that the information and regulations therein comply with the latest airworthiness standards and requirements and can support the safe operation of the aircraft. These manuals and materials are important bases for aircraft operation and maintenance. If the information is inaccurate or outdated, it may affect the safety of the aircraft.

② 损伤尺寸鉴定：对损伤进行评估，包括确定损伤的位置、尺寸、形状、类型、数量以及可能对飞机安全性和适航性产生的影响。

Damage size identification: Evaluates the damage, including determining the location, size, shape, type, quantity, and possible impact on the safety and airworthiness of the aircraft.

③ 损伤材料鉴定：检查损伤结构的材料可以确定需要使用何种类型的修复材料，检查厚度可以确定修理材料的合适厚度，检查状态可以帮助确定是否需要进行局部或整体更换。

Damage material identification: Checks the material of the damaged structure to determine the type of repair material needed, checks the thickness to determine the appropriate thickness of the repair material, and checks the status to help determine whether local or overall replacement is necessary.

④ 损伤评估：结合损伤评估结果，制定符合适航标准的修理方案。修理方案应考虑损伤的位置、尺寸、形状、类型、修理技术、材料和设备等因素，以及可能对飞机性能、强度、平衡和适航性产生的影响。

Damage assessment: Based on the results of the damage assessment, develop a repair plan that meets airworthiness standards. The repair plan should consider factors such as the location, size, shape, type of damage, repair technology, materials, equipment, and possible impact on aircraft performance, strength, balance, and airworthiness.

⑤ 编制修理方案：制定详细的修理方案，包括修理程序、材料规范、工具设备、检验方法、验收标准、适航文件等。修理方案应符合适航标准和制造商要求，确保修理后的结构满足设计要求和适航标准。

Preparation of repair plan: Develop a detailed repair plan, including repair procedures, material specifications, tooling, inspection methods, acceptance criteria, and airworthiness documents. The repair plan should comply with airworthiness standards and manufacturer requirements, and ensure that the repaired structure meets design requirements and airworthiness standards.

修理任务作为持续适航的一个环节，确定修理方案后，还需要进行实施和验收。实施阶段包括工具和材料的准备、修理的具体操作和流程的控制等。在修理过程中，还需要对维修任务进行记录和文档管理，确保记录的准确性和完整性。完成修理后，需要进行验收，确保修理的质量符合规定标准和适航要求。验收包括目视检查、结构强度测试、系统测试等环节，以确保修理后的部件或系统安全可靠，并符合适航标准。

As a part of the continuous airworthiness process, repairing a damaged aircraft involves not only determining the repair plan, but also implementing and verifying the repairs. The implementation phase involves preparing tools and materials, conducting the specific repair operations, and controlling the repair process. During the repair process, maintenance tasks should be recorded and documents managed to ensure the accuracy and completeness of records. After completing the repair, it is necessary to carry out acceptance tests to ensure that the quality of the repair meets the required standards and airworthiness requirements. Acceptance tests include visual inspections, structural strength tests, system tests, etc., to ensure that the repaired components or systems are safe, reliable, and compliant with airworthiness standards.

小任务3　按照损伤程度，结构修理是如何分类的？
Subtask 3　What is the classification of structural repair according to the degree of damage?

航空器结构的修理可分为三种类型：A 类、B 类、C 类。

Aircraft structure repair can be divided into three types: A, B and C.

（1）A 类修理：永久性修理、镶平修理，满足所有静强度要求与气动要求。

Class A repair: Permanent repair, flush repair, meet all static strength requirements and pneumatic surface requirements.

原区域检查程序的检查间隔及方法已经能够确保结构的持续适航性的结构修理方法。这就是说修理后的检查要求与原来的检查要求相同，不需给出补充结构检查，仍采用 MPD（维修计划数据）给出的维护检查间隔和方法。

The inspection intervals and methods of the original Zone inspection program have been used to ensure the continued seaworthiness of the structure. This means that the inspection requirements after repair are the same as the original inspection requirements. There is no need to provide supplementary structural inspection, and the maintenance inspection intervals and methods given by MPD (Maintenance Plan Data) are still adopted.

（2）B 类修理：过渡性修理、外部贴补修理，满足所有静强度要求，但抗疲劳和损伤容限细节不如 A 类镶平修理好。

Class B repairs: Transitional repair, external patch repairs that meet all static strength requirements but are not as good in fatigue and damage tolerance details as Class A trim repairs.

通常需要在使用一定的时间后，增加周期性的目视检查或无损检测的修理属于 B 类结构修理的范围。

Repairs that require periodic visual inspection or nondestructive testing after a certain period of use belong to class B structural repairs.

（3）C 类修理：临时性修理，仅满足静强度要求，不满足耐久性或气动要求。

Class C repair: Temporary repair, which only meets the requirements of static strength, but not durability or aerodynamic requirements.

这种修理通常是因为航空器停场时间短或修理条件不具备而采用。它需要比 B 类结构修理具有更严格的附加检查方法。可以按照 C 类结构维修手册提供的附加检查的门槛值和重复检查间隔进行检查，并应在规定期限内进行。

Such repairs are usually carried out because the aircraft has been parked for a short time or the repair conditions are not available. It requires additional inspection methods that are more stringent than those for class B structural repairs. Such class C structural repairs may be replaced within a specified period of time by inspection in accordance with the threshold values and repeat inspection intervals provided in the structural Repair manual for additional inspections.

小任务 4　飞机主要结构和次要结构的定义是什么？
Subtask 4　What is the definition of primary and secondary structure?

飞机结构分为主要结构和次要结构两大类。

Aircraft structures can be divided into two categories: Primary structure and secondary structure.

（1）主要结构：传递飞行、地面或者增压载荷的结构。主要结构包含重要结构（PSE/SSI) 和其他主要结构。

Primary structure: Structure for transferring flight, ground or pressurized loads. Primary structures include major structures (PSE/SSI) and other major structures.

（2）次要结构：仅传递局部气动载荷或者自身质量力载荷的结构。次要结构失效不影响结构持续适航性 / 飞行安全。大多数次要结构的主要作用为保证飞机气动外形、降低飞行时的空气阻力。

Secondary structure: The structure that only transmits local aerodynamic load or its own mass load. Secondary structural failure does not affect the continuous airworthiness/flight safety of the structure. The main function of most secondary structures is to ensure the aerodynamic shape of the aircraft and reduce the air resistance during flight.

小任务 5　修理定级的判定条款有哪些？
Subtask 5　What can be the repair classification check items?

根据航线维修施工规范，修理定级的判定条款主要包含如表 2-1-1 所示的相关项目。

According to the Airline Maintenance Construction Specification, the determining clauses for repair classification mainly include the relevant items as shown in Table 2-1-1.

表 2-1-1 修理定级判别依据

Table 2-1-1 Criteria for Determining Repair Classification

修理定级判别条款 Repair Classification Check Items	是 / Yes	否 / No
是否对飞机的以下适航特性有显著影响？（对适航性影响进行判别时应考虑的因素参照 DMDOR 工程手册） Does it have an appreciable effect on the following airworthiness characteristics of aircraft? (Refer to DMDOR manual for airworthiness concern on repair design classification) □质量 /Mass　　　　　□结构强度 /Structural Strength □平衡 /Balance　　　　□动力特性 /Power Plant Characteristic □性能 /Performance　　□飞行操作特性 /Flight Operation Characteristic □其他方面的适航特性 /Other Airworthiness Aspects		
是否需要进行复杂的静力、疲劳及损伤容限分析或测试？（需要分阶段批准的修理视为符合该规则） Are extensive static, fatigue and damage tolerance strength justification and/or testing require? (Repairs that require stage-approval is considered as typical case of this condition)		
是否需要对飞机原有的审定认证数据进行重新评估或审定以确保飞机满足所有相关要求？ Are the re-assessment and re-evaluation of the original certification substantiation data require to ensure that the aircraft still complies with all the relevant requirements?		
是否对寿命有限件或关键部件进行加强修理？ Is it necessary reinforcing repair to life limited or critical parts?		
修理的实施是否需要通过非常规做法或非常规作业（如选择非常规材料、热处理、材料工艺流程、夹具定位图等）？ Does the implementation of repairs require unconventional practices or unconventional operations (such as choosing unconventional materials, heat treatment, material processing methods, jigging diagrams, etc.)?		
是否引起飞行手册 (AFM) 的改变？ Did it cause a change to Aircraft Flight Manual (AFM)?		
是否引起飞机维修方案 (AMS) 的改变？（具体见 DMDOR 工程手册对该条款的解释） Did it cause a change to Approved Maintenance Schedule(AFM)? (See DMDOR manual for detail explanation)		
修理分类为 /Repair is classified as: □重要修理 /Major　　□非重要修理 /Minor 注释：若以上所有条款评估结果皆为"否"，则该修理将判定为"非重要修理"，否则判定为"重要修理"。 Note: If 'No' in all the above items, the repair shall be classified as 'Minor', otherwise as 'Major'.		

小任务 6　仅参考 SRM，是否足以解决所有维修任务？

Subtask 6　Is it enough to only refer to the SRM to solve all maintenance tasks?

仅参考 SRM 无法解决所有维修任务。尽管 SRM 是一个重要的参考文献，但它并不能涵盖所有可能出现的维修任务。对于一些复杂的维修任务，需要更多的信息和专业知识来制定维修方案。另外，SRM 也可能不包括最新的维修技术和方法，因此维修人员可能需要进行其他方面的研究和学习。在处理超手册结构修理工作时，更需要适航部门的批准和指导，以确保维修任务的安全和适航性。

Relying solely on SRM may not address all maintenance tasks. Although the SRM is an important reference document, it cannot cover all possible maintenance tasks that may arise. For some complex maintenance tasks, more information and professional knowledge are required to develop a maintenance plan. In addition, the SRM may not include the latest maintenance techniques and methods, so maintenance personnel may need to conduct additional research and learning. When dealing with major repair tasks, it is even more important to obtain approval and guidance from the airworthiness authority to ensure the safety and airworthiness of the maintenance task.

随着适航观念的日益深入，适航部门包括各个航空公司对飞机结构修理批准越来越重视，所有存在于飞机上的修理，无论重要修理还是一般修理，如果超出持续适航文件的范围，都需要得到适航部门的批准或认可。

As the concept of airworthiness deepens, not only the airworthiness department including various airlines pays more and more attention to the approval of aircraft structural repairs. All repairs on the aircraft, whether they are major or minor, require approval or recognition from the airworthiness department if they exceed the scope of continuous airworthiness documents.

通过查询 SRM 能解决多数常见的"小"损伤，而实际维修任务中涉及飞机结构完整性损伤层面的修理，称为"超手册结构修理"，或者称为"重要修理"（Major Repair）。相较于一般修理（Minor Repair），重要修理是指如果不正确地实施，将可能导致对质量、平衡、结构强度、性能、动力特性、飞行特性和其他适航性因素有明显影响的修理。重要修理不是按照已经被接受的方法或者通过基本的作业就能够完成的工作。对于不同类型的超手册结构修理，如何快速地得到飞机制造国和注册国适航部门的批准，是一个较为复杂的过程。

Most common "small" damages can be solved by querying the SRM. However, for repairs involving aircraft structural integrity damage, they are called "beyond manual structural repairs" or "major repairs". Compared with minor repairs, major repairs refer to repairs that, if not implemented correctly, may have a significant impact on weight, balance, structural strength, performance, dynamic characteristics, flight characteristics, and other airworthiness factors. Major repairs are not jobs that can be completed by accepted methods or basic operations. For different types of beyond manual structural repairs, obtaining approval from the airworthiness departments of the aircraft manufacturing and registration countries quickly is a complex airworthiness process.

表 2-1-2 给出发动机吊架面板损伤修理方案过程，供制定超手册修理方案的技术人员和结构工程师参考。

Table 2-1-2 presents the process for repairing engine mount panel damage, serving as a reference for technical personnel and structural engineers in developing repair procedures for beyond manual structural repairs.

表 2-1-2　超手册修理案例——发动机吊架面板损伤

Table 2-1-2　Repair beyond Specifications Example—Pylon Panel Found Damaged

技术协助申请 TS	申请号：××000000
主题：发动机第二吊挂面板发现损伤 Subject: NO. 2 PYLON PANEL FOUND DAMAGED	TC NO:××000000
缺陷描述 /Damage Description NRC 发动机第二吊挂面板发现损伤，面板材质为塑料，没有相关维修文件，请协助处理。 NRC—NO. 2 ENG PYLON PANEL WAS FOUND DAMAGED. BUT THE PANEL MADE FROM PLASTIC HAS NO RELATED REPAIR DOCUMENT, PLEASE ADVISE.	
技术服务部回答 / TSD REPLY： 请按附件 RSK-XX-XXXX REV. 00 进行维修。 PLEASE CARRY OUT REPAIR AS PER ATTACHED RSK-XX-XXXX REV. 00.	
改装设计委任单位代表工程文件 / 资料批准表 DMDOR Approval of Engineering Document & Design Data Package	

DMDOR 修理方案： RSK-XX-XXXX 版次号 00 DMDOR Repair Scheme: RSK-XX-XXXX Rev.00 参考 Reference： Boeing 服务请求号：X-000000 Boeing SR No.: X-0000000000	2 号发动机反推和吊架上前方内侧整流面板边缘裂纹。 Crack C/T#2 Engine T/R to Strut Forward Up INBD Fairing Panel Edge. 该修理划分为次要时限性修理，所需要的检查和限制要求显示在 RSK-XX-XXXX Rev.00 This repair is classified as a MINOR Time 0-Limited repair with inspection and limitation in RSK-XX-XXXX Rev. 00.

资料的目的 Purpose of Data：	
□STC 改装 STC Modification	□重要改装 Major Modification
□非重要改装 Minor Modification	□超规范修理 Repair beyond Specifications

DMDOR 修理方案 Repair Scheme

修理描述 /Description of Repair：
2 号发动机反推和吊架前方上部内测整流面板（件号：000-0）边缘发现裂纹。该裂纹末端已钻出止裂孔，且裂纹区域填充树脂并层叠玻璃纤维布使用胶液进行固化。

#2 engine T/R to strut forward upper INBD fairing panel (P/N: 000-0) edge was found cracked. This crack damage was stop drilled at the end and the cracked area was filled with resin and cured by laying up fiberglass and adhesive.

执行步骤 Accomplishment Instructions

（1）清洁损伤区域的污渍、油渍和油脂。
Clean damaged area to remove all dirt, oil and grease.
（2）对整个维修区域进行打磨。以去除油漆、底漆和 / 或填充片。
Sand entire repair area. Remove paint, primer and/or filler.

执行步骤 Accomplishment Instructions

（3）在裂纹末端钻出一个直径 3/16 in 的止裂孔。

Stop drill crack with 3/16 inch diameter holes.

（4）根据 xxx-x SRM 51-70-06 清洁维修区域，用溶剂清洗裂纹表面，使用刷子涂刷一层充足的树脂涂层。密封裂纹区域并且保持密封直到树脂固化。

Clean the repair area as per ×××-× SRM 51-70-06, after solvent cleaning the crack surfaces, apply a generous coat of resin system on the cracks by brush. Close the crack and hold closed until resin is cured.

注释：确保止裂孔完全被树脂混合填充并且两侧完全被布层覆盖。

Note: Ensure that the stop drill holes are completely filled with resin mix and covered on both sides by the fabric plies.

（5）清洗干净所有打磨产生的粉尘。

Clean off all dust generated from sanding.

注意：粉尘会导致蒙皮应激 / 吸附效应。建议使用蒙皮吸附保护。

Caution: Dust can cause skin irritation/breathing complications. Advise use of skin espiration protection.

（6）根据 xxx-x SRM 51-70-06 在裂纹区域铺三层布：铺第一层玻璃纤维布，使布在外表面从裂纹边缘线向外延伸 2.5 in 并且在内表面从裂纹边缘线向外延伸大约 1.5 in；铺第二层玻璃纤维布，使布在外表面从裂纹边缘线向外延伸 3.0 in 并且在内表面从裂纹边缘线向外延伸大约 2.0 in；铺第三层玻璃纤维布，使布在外表面从裂纹边缘线向外延伸 3.5 in 并且在内表面从裂纹边缘线向外延伸大约 2.5 in。

Install three repair plies to the cracked area as follows and per xxx-x SRM 51-70-06: Install the first fiberglass ply extending 2.5 inches above the crack-line on the outer surface and approximately 1.5 inches on the inner surface; Install the second ply extending 3.0 inches on the outer surface and approximately 2.0 inch on the inner surface; The third ply shall extend 3.5 inch on the outer surface and approximately 2.5 inch on the inner surface.

注释：玻璃纤维维修布层为 Type 0000。

Note: Glass fabric repair plies Type 0000.

（7）对以上描述的维修执行双阶段固化。

Perform the repair with a two-stage cure.

（8）在固化期间使用铝片或者等效方法来夹紧修补区域。

Clamp repair between two aluminum plates, or equivalent, during cure.

注释：制作两块铝板或使用等效方法，固化时施加均匀的压力在维修区域。在铝板或者等效的夹紧装置和修补区域接触的侧面施加聚酯纤维胶层或者合适的脱模剂。

Note: Fabricate two aluminum plates, or equivalent, to apply even pressure over the repair area during cure. Apply polyester film tape or appropriate release agent to the repair side of the plates or equivalent clamping device.

（9）移除所有的夹具和夹紧铝片或者等效装置。

Remove all clamps and clamping plates or equivalent.

（10）根据 AMM 51-21-00 恢复表面处理。

Restore finish according to AMM 51-21-00.

通用注释：

General note:

（1）零件表面需打磨和使用溶剂擦洗。

The part shall be surface prepped by solvent wipe, abrading and solvent wipe.

（2）零件需在 250 ℉温度下干燥 1~2 h，以减少固化维修过程中水分的溢出。

The part shall be dried 1 to 2 hours at 250 ℉ prior to repair in order to reduce the release of water during the cure of the repair.

（3）固化时，真空袋可代替铝板施加压力在修理铺层上。

Vacuum bag is an alternative to using aluminum plates to apply pressure on lay up during cure.

手册查询练习任务工卡
Manual Training Task Card

查询 B787-SRM 手册，完成以下练习。

（1）下列哪项不属于"Principal Structural Elements"？（　　　）

       A. Main Landing Gear Axles   B. Main Landing Gear Hanger Link

       C. Cargo Floor Panels   D. Aileron Hinges

（2）SRM 手册的 53 章为 (　　　)。

（3）在 SRM 手册 51 章中，标题为 Repairs 的章节号为 (　　　)。

（4）在 SRM 手册 51 章中，章节号为 51-30 的标题为 (　　　)。

（5）Stabilizers 属于哪一章？（　　　）

（6）51-40 章节对应的内容是 (　　　)。

       A. Damage Evaluation and Removal-Metals   B. Fasteners

       C. Composite Repair-Common Data   D. Repairs

（7）53-10-01 章节号对应的内容是 (　　　)。

       A. Fuselage-GENGEL-Skin   B. Fuselage-Section 41-Skin

       C. Fuselage-Section 41-Stringers   D. Fuselage-Section 44-Skin

（8）Serial Number 是指 (　　　)。

       A. 机型   B. 生产线号

       C. 序列号   D. 注册号

（9）对于同一架飞机来说，下列号码唯一的是 (　　　)。

       A. Model-Series   B. Identification Code

       C. Effectivity Code   D. Serial Number

（10）下列说法错误的是 (　　　)。

       A. 前货舱门 Zone: 821   B. 后货舱门 Zone: 821

       C. 前货舱门前缘 STA: 441   D. 后货舱门前缘 STA: 441

子任务 2　ADL 查询
Subtask 2　ADL Consultation

【子任务解析 Subtask Analysis】

可允许损伤限制又称为损伤容限（Allowable Damage Limit，ADL），是决定构件剩余强度与寿命要求关系的一个重要指标，是损伤定级与确定修理方案的重要依据，如果损伤在 ADL 以内，那么该损伤不需要永久性修理，只需要密封处理即可。确定 ADL 的过程通常为大量复杂的有限元分析和模拟分析，在本书中不作详细探讨。

ADL (Allowable Damage Limit), an important parameter to determine the residual strength and life requirements, is also essential for damage classification and repair plan determination. If the damage is within ADL, permanent repair is not required while a seal repair is recommended. The process of determining ADL usually involves a lot of complex finite element analysis and simulation analysis, which is not discussed in detail in this book.

通过本任务的训练，可掌握在 SRM 查找 ADL 的基本方法，看懂 ADL 相关表、图和说明，以了解修理方案的制定依据，从而准确实施不同损伤类型、不同损伤尺寸的修理方案。

By completing this training task, one can master the basic method of searching for ADL in the SRM, understand the tables, figures, and instructions related to ADL, and have a deep understanding of damage classification. This will enable accurate implementation of repair plans for different types and sizes of damage.

【子任务分解 Subtask Break-down】

（1）理解 ADL 的定义。

Know the definition of ADL.

（2）基本掌握 ADL 查找方法。

Master the method of searching ADL.

（3）理解 ADL 相关图、表和说明。

Understand related diagrams, tables and instructions.

小任务 1　什么是 ADL？ ADL 与 RDL（Repairable damage limit）的关系是什么？

Subtask 1　What is the ADL? What is the relationship between ADL and RDL?

可允许损伤定义为不影响结构完整性或降低构件功能的轻微损伤。

Allowable damage is defined as a minor damage which does not affect the structural integrity or decrease the function of a component.

ADL 来源于结构设计理论，该理论假设结构的疲劳失效是从微缺陷萌生开始的，而这些缺陷的扩展速度以及结构的剩余强度便是影响构件寿命的重要因素，损伤容限便是用来描述结构在服役过程中，抵抗由缺陷、裂纹或其他损伤而导致破坏的能力。

Damage tolerance is derived from the theory of structure design, the fatigue failure of the structure of the theory assumes that starts from the minor defect initiation, and these defects extension speed and the structure of the residual strength is one of the important factors affecting the service life of components, and damage tolerance is used to describe the structure in the process of service, the resistance caused by defects, cracks or other damage and destruction.

构件出现微缺陷，即初始损伤出现，为损伤容限的下限；构件发生断裂破坏，为损伤容限的上限。

The lower limit of damage tolerance is the occurrence of initial damage when the component has microdefect. Fracture failure of component is the upper limit of damage tolerance.

修理容限（RDL）是指在安全因素与修理验收指标规定下，为"需要与不需要修理""能与不能修理"的两个定量进行限界的参数，而且修理容限通常小于损伤容限。

Repair tolerance refers to the parameters that limit the two quantitative values of "need or not need repair" and "can or can't repair" under the provisions of safety factors and repair acceptance indexes, and the repair tolerance is usually less than the damage tolerance.

根据损伤尺寸所在的"容限区间"，分为允许损伤、可修理损伤和不可修理损伤，如图 2-2-1 所示。

Damage is classified into allowable damage, repairable damage, and unrepairable damage according to the "tolerance range" of the damage size, as shown in Fig.2-2-1.

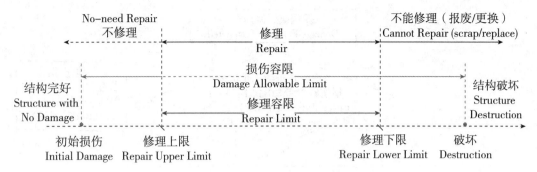

图 2-2-1　损伤容限与修理容限

Fig.2-2-1　ADL VS RDL

小任务 2　确定某损伤是否为允许损伤的步骤是什么？
Subtask 2　What are the steps to determine if a damage is ADL?

整体流程如图 2-2-2 所示，具体步骤说明为：

The overall process is shown in Fig.2-2-2, with detailed step-by-step instructions as follows:

（1）在 SRM 内找到损伤结构的相关章节。

Find the SRM Chapter and Section for the Damaged Structure.

（2）查找允许损伤限制 (ADL)。

Find the Allowable Damage Limits (ADL).

（3）查找允许损伤限制 (ADL) 的适用性。

Find the ADL applicability.

（4）查找需要的损伤检查方法。

Find the required damage inspection method.

例如，对于 43 段机身蒙皮的损伤检查，通用程序会显示需要目视检查，大多数情况下只需要目视检查。

As an example: For fuselage section 43 skin damage, general routine shows visual inspection. Most allowable damage limits only required visual inspection.

（5）确定损伤尺寸。

Find the damage dimensions.

（6）确认损伤类型。

Find the type of damage.

（7）在允许损伤限制内查找适用的表、图和说明等。需要将所有适用的表、图和说明找出。

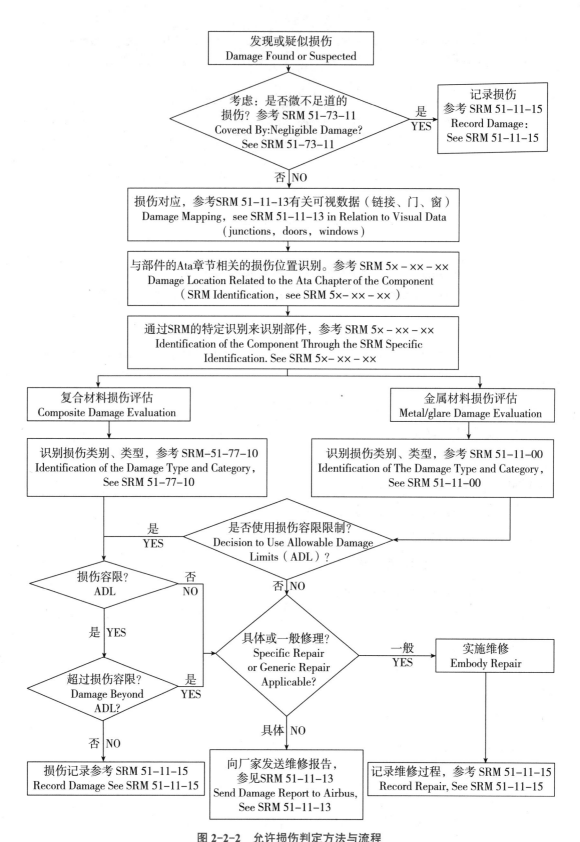

图 2-2-2 允许损伤判定方法与流程
Fig.2-2-2 ADL classification flow chart

Find the applicable ADL figures, tables, and notes for the damage. After you find the damage type, you must find all ADL table notes that apply.

（8）检查损伤是否在允许损伤限制内。

Find if the damage is within the allowable damage limits.

将损伤的尺寸（所有数据）与允许损伤限制内数据比较。

Compare the dimensions of the damage (all data) with the allowable damage limits.

（9）查找飞机的放行要求，飞机放行前，损伤应密封处理。必须对损伤部件进行密封处理，防止树脂受潮气和紫外线的侵害。按 SRM 要求进行密封，如用树脂密封或用快速胶带密封。

Find the dispatch requirements for the permitted damage. Damage must be sealed before dispatch. You must seal the part to protect the resin from moisture and ultra violet radiation. Examples of sealing methods are resin or speed tape.

如果确认损伤为允许损伤，只需要密封处理即可。

If the damage dimensions are less than the allowable damage limits, only sealing is necessary.

思考：对于 43 段机身蒙皮，发现紧固件位置存在约 3 in 的裂纹，请问是否超过损伤容限？

Thinking: For the 43 section fuselage skin, a crack of about 3 in. is found at the fastener location. Does it exceed the ADL?

维修方案制定练习题任务工卡

Maintenance Scheme Determination Exercise Task Card

（1）制定复合材料维修方案时需要考虑哪些因素？这些因素与金属材料的因素有何不同之处？

What are the factors that should be considered when developing a maintenance plan for composite materials? How do these factors differ from those for metallic materials?

（2）举例描述复合材料结构损伤的维修方案制定步骤，并画出相应修理方案流程图。

Provide an illustrative description of the steps involved in formulating a repair plan for composite structural damage, and create a corresponding flowchart for the repair procedure.

任务 2　维修方案制定考核表

姓名		班级		
评价维度	分值/分	自评（30%）	互评（30%）	师评（40%）
素养（20%）				
1. 手册自学情况	5			
2. 任务书自学情况	5			
3. 安全文明操作及 6S	10			
技能（60%）				
1. 能查阅正确章节	30			
2. 能查阅施工程序	15			
3. 能制定维修方案	15			
总结报告（20%）				
1. 训练总结的完整性	10			
2. 个人反思与拟订后续训练计划	10			
总计	100			
任务完成情况	提前完成			
	准时完成			
	滞后完成			

去除损伤与型面准备
Damage Removal & Surface Preparation

【任务情境 Task Scenario】

您是 WS 航空公司的航线结构修理部门操作员,并持有复合材料结构件修理授权书。某日你到岗后,收到航线报机身右侧壁板存在凹坑损伤,深度超过可允许损伤容限,需进行永久性修理。你的工作是根据超声检测结果,按照手册要求完成台阶打磨并确保去除全部损伤区域。

You are an operator in the Line Structural Repair department of WS Airlines, holding an authorization for the repair of composite structural components. One day, when you arrived at work, you received a report that the right side wall panel of the aircraft had a dent and damage deeper than the allowable damage tolerance, requiring a permanent repair. Your job is to perform step-by-step polishing according to the manual requirements and ensure that the entire damaged area is removed based on the results of ultrasonic testing.

【任务解析 Task Analysis】

去除损伤是进行复合材料结构损伤维修工作的第三步。在进行复合材料结构损伤维修工作时,必须确保受损区域的表面清洁、光滑,以确保修复材料可以黏附并达到最佳效果。

Removing damage is the third step in repairing damaged composite structures. When carrying out repair work on composite structures, it is essential to ensure that the damaged area's surface is clean and smooth to ensure that the repair material can adhere and achieve the best results.

完成本任务训练后,应该能够实现以下目标。

After completing training for this task, the following goals should be achieved.

知识目标 Knowledge Objectives

（1）熟悉去除层合板及蜂窝夹层板零件的损伤层的标准程序。

Be familiar with standard procedures for removing damage layer of laminates and honeycomb sandwich panels.

（2）理解安全操作规程：学习使用工具和材料时的安全操作规程，以确保在操作过程中的人身安全和工作环境安全。

Understanding of safety operating procedures: Learn the safety operating procedures when using tools and materials to ensure personal and work environment safety during operation.

能力目标 Ability Objectives

（1）按照相应机型的结构维修手册（SRM）等文件，正确地进行零件保护、零件清洁与干燥以及规范使用去除损伤的工具和进行人员的安全防护。

According to the corresponding machine structure repair manual (SRM) and other documents, correct parts protection, parts cleaning and drying, standard use of damage removal tools and safety protection of personnel.

（2）熟练使用去除损伤的工具：能够熟练地使用各种手动和机械工具，如打磨机、镂铣机等。

Proficient use of damage removal tools: Be able to proficiently use various manual and mechanical tools, such as grinding machines, milling machines, etc. （3）熟练使用型面准备工具：能够熟练地使用各种型面准备工具，如清洁剂等，准备受损区域的表面。

Proficient use of surface preparation tools: Be able to proficiently use various surface preparation tools, such as cleaning agents, to prepare the damaged area's surface.

素质目标 Emotion Objectives

（1）机务作风与安全意识：确保个人安全和工作环境的安全至关重要。维修人员应熟悉安全规范，包括穿戴适当的个人防护装备（PPE）和安全使用工具以防止受伤。

Maintenance work and safety awareness. Ensuring personal safety and the safety of the work environment is crucial. Repair personnel should be well-versed in safety protocols, including wearing appropriate personal protective equipment (PPE) and using safety tools to prevent injury.

（2）具有细心和耐心的品质，能够对复合材料结构的损伤区域进行精确和仔细的观察和评估，确保去除损伤和型面准备工作的质量。

The qualities required for this task include meticulousness and patience to accurately and carefully observe and evaluate the damaged area of the composite structure, ensuring the quality of the damage removal and surface preparation work.

（3）团队合作与沟通能力：学会与团队成员配合，共同完成检修任务。培养与客户、同事、监察有效沟通的能力。

Team cooperation and communication skills: learn to cooperate with team members to complete maintenance tasks. Develop the ability to communicate effectively with customers, colleagues, and supervisors.

【任务分解 Task Break-down】

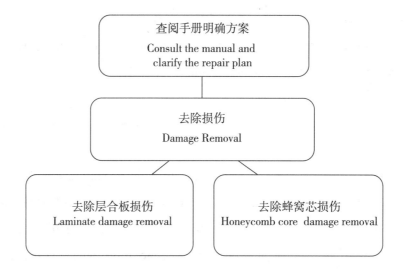

子任务 1　清洁与干燥
Subtask 1　Cleanup and Absorbed Moisture Removal

【子任务解析 Subtask Analysis】

在实施去除损伤之前，必须明确维修方案，了解此零件制造图纸，观察零件是否污染，判断零件是否侵入水汽等，因此需要进行相应的准备工作，如表面清洁、干燥等工序。

Before the removal of damage, it is necessary to clarify the maintenance plan, understand the manufacturing drawings of the parts, observe whether the parts are contaminated, determine whether the parts are invaded by water vapour, so it is necessary to carry out the corresponding preparation work, such as surface cleaning, drying and other processes.

通过本任务的训练，可掌握常用层合板以及蜂窝夹层板零件去除损伤的去漆层、清洁与干燥工序。

Through the training of this task, competitors can master the process of removing paint layer, cleaning and drying of commonly used laminates and honeycomb sandwich panels.

【子任务分解 Subtask Break-down】

（1）掌握表面清洁工序。

Master the surface cleaning process.

（2）掌握快速干燥方法。

Master rapid drying methods.

【背景知识 Background Knowledge 】

小任务 1　为什么不能直接去除损伤？
Subtask 1　Why can not the damage be directly removed?

明确维修方案与零件制造图纸是去除损伤前的重要一步。不能直接去除损伤主要有以下四点需要考虑：

Defining the repair plan and part manufacturing drawings is an important step before removing damage. The following four points should be considered why the damage cannot be removed directly:

（1）贸然去除损伤可能与维修方案发生冲突造成工作重复浪费，甚至会对零件造成二次伤害而无法执行后续维修任务；

The reckless removal of damage may conflict with the maintenance plan, resulting in repeated waste of work, or even secondary damage to parts and unable to perform follow-up maintenance tasks;

（2）不了解零件结构制造图纸，无法根据铺层数和铺层角度来辅助判断去除深度；

Do not understand the part structure manufacturing drawings, can not be based on the number of layers and layer angle to assist in judging the removal depth;

（3）零件表面有油等污染，直接去除损伤可能导致污染进入零件内部不利于粘接；

There is oil and other pollution on the surface of the parts, direct removal of damage may lead to pollution entering the parts, which is not conducive to bonding;

（4）零件内部水侵入，如蜂窝积水，不利于或者无法进行打磨等维修工序。

Parts internal water intrusion, such as honeycomb water, is not conducive to or cannot be polished and other maintenance processes.

扩展 1　复合材料零件结构形式主要有层合板结构与夹芯结构两种形式。

Extension 1　The structural forms of composite parts are mainly laminated and sandwich structures.

在飞机结构中，复合材料夹芯结构的面板，复合材料翼盒结构的梁、肋、壁板和机身侧壁板等通常采用层合板结构。层合板结构应用于承载环境较为简单的结构，通常为承受轴向或对抗冲击性能要求较低的构件。操纵舵面（副翼、升降舵和方向舵）、扰流板、雷达罩和客舱底板等通常采用夹芯结构。

In aircraft construction, laminate structures are commonly used for panels of composite sandwich structures, beams, ribs, panels and fuselage side panels of composite wing box structures. Laminate structures are used for structures with relatively simple load bearing environments, usually members with low requirements for axial or anti-impact performance. The control surfaces (ailerons, elevators, and rudder), spoilers, radomes, and cabin floor are usually of sandwich construction.

层合板又称层压板，是由若干层湿铺层或预浸料铺层按照某种铺层设计以铺叠粘接的形式，经过加温加压、固化而成的多层板材（图3-1-1）。其中湿铺层是指在工作现场，用树脂将干的纤维织布或纤维浸渍后所形成的补片铺层。

Laminate is composed of several layers of wet layering or prepreg layering in accordance with a kind of layering design in the form of stacking bonding, after heating and pressure, curing multilayer plate(Fig.3-1-1). The wet layer refers to the patch layer formed after the dry fiber is woven with resin or the fiber is impregnated at the work site.

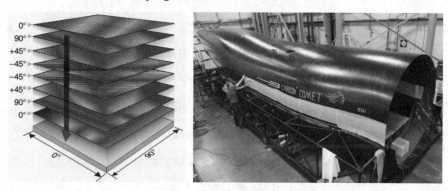

图3-1-1 层合板示意图与实物图
Fig.3-1-1 Schematic and physical drawings of laminates

复合材料夹芯结构是由上、下面板与芯材用胶粘剂粘接而成的整体结构。面板通常是复合材料层合板。芯材有蜂窝芯、泡沫芯和波纹板芯等多种形式，其中用得最多的是蜂窝芯。蜂窝芯有金属蜂窝芯和非金属蜂窝芯两大类。金属蜂窝芯主要是铝蜂窝芯材，非金属蜂窝芯材有玻璃布蜂窝芯材、纸蜂窝芯材和Nomex蜂窝芯材等（图3-1-2）。

The composite sandwich structure is an integral structure composed of the upper and lower panels bonded with the core material by an adhesive. The panel is usually a composite laminate. Core materials have honeycomb core, foam core and corrugated core and other forms, among which the most used is honeycomb core. Honeycomb core has two categories: metal honeycomb core and non-metal honeycomb core. Metal honeycomb core is mainly aluminum honeycomb core material, non-metal honeycomb core material has glass cloth honeycomb core material, paper honeycomb core material and Nomex honeycomb core material (Fig.3-1-2).

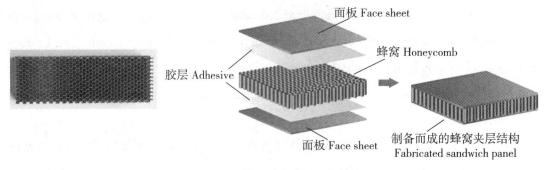

图 3-1-2　蜂窝夹层板实物图与示意图

Fig.3-1-2　Honeycomb structure composite physical and schematic images

扩展 2　湿气的侵入问题是造成结构失效以及航空安全隐患的重要一题。

Extension 2　Moisture intrusion is an important problem that causes structural failure and aviation safety risks.

以雷达罩为例,雷达罩对雷达天线的作用堪比风挡对飞行员的作用。一旦雷达罩萌生裂纹甚至存在明显结构损伤,外层的大气和夹层蜂窝的内部空间就会相通,湿气的侵入将导致雷达罩局部进水。即使水分侵入雷达罩的含量极少,也会影响雷达罩的传输效率,甚至导致雷达天线的辐射图和发射信号失真。因此,气象雷达罩的受损积水情况直接关系着航空运输业的安全,这是一个绝对不容忽视的安全问题。

Radome, for example, are as useful to a radar antenna as windshields are to a pilot. Once cracks or even obvious structural damage occur in the radome, the outer atmosphere will communicate with the internal space of the sandwich honeycomb, and the invasion of moisture will lead to local water intake in the radome. Even a small amount of moisture can affect the transmission efficiency of the radome and even lead to distortion of the radiation pattern and transmitted signal of the radar antenna. Therefore, the damaged water of weather radome is directly related to the safety of air transport industry, which is an absolute safety problem that cannot be ignored.

小任务 2　熟悉制造图纸中铺层角度的表示
Subtask 2　Acknowledge the representation of laying sequence method in manufacturing drawings

一般常采用公式表示法,即将全部铺层以铺层角数字的形式写入中括号"[ ]"内,各铺层按由下向上或贴膜面向外的顺序写出,各铺层的铺层角数字用"/"分开,同时根据需要标以下脚标、顶标等,以表达各铺层数、总铺层数、铺层材料和铺层顺序等,如表 3-1-1 所示。

General formula is often used to representation, that is, the whole angle of layer by layer numbers in the form of a written inside the parentheses "[ ]", each layer from the bottom up or sticker and write to order, each layer of the angle of layer numbers separated by "/", at the same time according to need standard feet below standard, standard, etc., to express the layer number, the total number of layer, layer and layer material order, etc., as shown in Table 3-1-1.

表 3-1-1　复合材料图纸铺层角度表达方式

Table 3-1-1　Representation of laying sequence method for composite

分类 Classification		图示表达法 Graphical representation method	公式表达法 Formula expression method	说明 Description
一般层合板 General laminates		90 −45 0 45 0	[0/45/0/−45/90]	每一铺层的方向用铺层角表示 The direction of each ply is represented by the ply angle.
对称 层合板 Symmetric laminates	偶数层	0 90 90 0	[0/90]s	每一铺层的方向用铺层角表示。彼此用"/"分开、全部铺层用"[ ]"示出。 铺层按由下向上或由贴膜面向外的顺序写出 The orientation of each ply is indicated by the ply angle. They are separated by "/" and all plies are enclosed within "[ ]", written in the order from bottom to top or from the film side outward.
	奇数层	0 45 90 45 0	[0/45/90]s	对称铺层的奇数层层合板, 在对称中间的铺层上加顶标"—", 其余和偶数层相同 For symmetrical ply stacking, a dashed line ("—") is added on the ply in the middle of the symmetry, while the others remain the same as even plies.
由多个子层合板构成的层合板 A laminated panel composed of multiple sub-layers		90 0 90 0 90 0	[0/90]3	子层合板重复数用下标示出 Sublayer laminate repetition numbers are indicated using subscripts.
织物组成的层合板 A fabric-reinforced composite laminate		0, 90 ± 45	[(±45)/(0,90)]	织物的经纬方向用"( )"示出 The warp and weft directions of the fabric are indicated using parentheses.

73

分类 Classification	图示表达法 Graphical representation method	公式表达法 Formula expression method	说明 Description
混杂纤维层合板 A hybrid fiber composite laminate	90_C 45_K 0_C	$[0_C/45_K/90_G]$	各纤维种类用英文字母下标示出： C- 碳纤维，K- 芳纶，G- 玻璃纤维 Different types of fibers are indicated using English letter subscripts: C – carbon fiber, K – aramid fiber, G – glass fiber.
夹芯板 Sandwich panel	0 90 C_4 90 0	$[0/90/C_4]s$	用 C 表示夹芯，下标数字表示夹芯厚度的毫米数 "C" represents the core material, and the subscript number indicates the thickness of the core in millimeters.

不同的铺层角度设计由不同的受力载荷决定，[0/+45/90/−45] 为通过试验验证的载荷最大铺层设计，其余交付设计如表 3-1-2 所示。

Different ply laying sequence designs are determined by different load requirements. [0/+45/90/−45] is the ply design with the highest load capacity verified by experiments, and other designs are presented in Table 3–1–2.

表 3-1-2 铺层设计案例
Table 3–1–2 Laying sequence design examples

受力性质 Load form	层合板结构形式 Structural laying sequence	用途 Application
承受拉伸、压缩载荷，可承受有限的剪切载荷 Capable of enduring both tensile and compressive loads while maintaining resistance to limited shear forces.	0/90/90/0 或 90/0/0/90	主要应力状态为拉伸应力或压缩应力，或拉、压双向应力的构件设计 The primary stress state involves either tensile stress or compressive stress, or bidirectional stress of both tension and compression in the component design.
承受拉伸载荷、剪切载荷 Capable of withstanding tensile and shear loads.	45/−45/−45/45 或 −45/45/45/−45	用于主要应力为剪切应力的构件设计 Utilized for component designs primarily subjected to shear stress.
承受拉伸载荷、压缩载荷、剪切载荷 Capable of withstanding tensile, compressive, and shear loads.	0/45/90/−45/−45/90/45/0	用于主要应力为剪切应力的构件设计 Utilized for component designs primarily subjected to shear stress.
承受压缩载荷、剪切载荷 Capable of withstanding compressive and shear loads.	45/90/−45/−45/−0/45	用于压缩应力和剪切应力，而剪切应力为主要应力的构件设计 Utilized for component designs subjected to compressive and shear stresses, with shear stress being the primary stress.

受力性质 Load form	层合板结构形式 Structural laying sequence	用途 Application
承受拉伸载荷、剪切载荷 Capable of withstanding tensile and shear loads.	45/0/−45/−45/0/45	用于拉伸应力和剪切应力，而剪切应力为主要应力的构件设计 Utilized for component designs subjected to tensile and shear stresses, with shear stress being the primary stress.

铺层设计案例 A laying-up sequence design example

假设需要设计一块尺寸为 1 m×1 m 的复合材料板，用于承受在 X 方向上的拉伸载荷。使用以下材料：

Assuming that a composite material panel with dimensions of 1 m×1 m is required to withstand tensile loads in the X direction. The following materials are used:

（1）碳纤维双向织物（0°，90°）Carbon fiber bidirectional fabric (0°，90°)

（2）碳纤维单向布（0°）Carbon fiber unidirectional fabric (0°)

（3）碳纤维单向布（45°）Carbon fiber unidirectional fabric（45°）

为了使复合材料板在 X 方向上具有最大强度，可以将铺层设计为 [0/90/0/45/−45/0/90]。其中，0 表示碳纤维单向布（沿着 X 轴方向），45 表示碳纤维单向布（沿着 45°对角线方向），−45 表示 45°对角线的相反方向。

To maximize the strength of the composite panel in the X direction, the layering can be designed as [0/90/0/45/−45/0/90]. Here, 0 represents carbon fiber unidirectional fabric (along the X-axis direction), 45 represents carbon fiber unidirectional fabric (along the 45 degree diagonal direction), and −45 represents the opposite direction of the 45 degree.

小任务3　表面清洁需要注意什么？
Subtask 3　What should be noted for surface cleaning?

表面清洁可能在维修工作当中操作多次，每次目的不尽相同，此表面清洁工序是在去除损伤工作之前实施，主要针对表面污染清理，利于开展去除损伤工作。表面清洁通常会使用甲基乙基酮（MEK）、丙酮或三氯甲烷等有机溶剂以及无绒干净的抹布或者擦拭纸。

Surface cleaning may operate many times in the maintenance work, each time the purpose is not the same. As for this task, the surface cleaning process is implemented before the removal of damage work, mainly for surface pollution cleaning to prepare the removal of damage work. Surface cleaning is usually done with solvents such as Methyl Ethyl Ketone (MEK), acetone, or chloroform combined with a lint-free rag or wiping paper.

指定标准大修操作手册 (SOPM) 作为清除污染的清洁步骤的参考：在清除损伤之前，在进行黏合程序之前，在进行保护性表面处理之前。SOPM 包含了这些清洗步骤中允许使用的溶剂清单。每一步都有一个不同的列表。甲基乙基酮 (MEK)、溶剂、B00148 在所有 SOPM 参考中用于清洗复合结构。

The Standard Overhaul Practices Manual (SOPM) is specified as a reference for a cleaning step to remove contamination: Before you remove damage, before a bond procedure, and before applying a protective finish. The SOPM contains a list of solvents that are permitted for each of these cleaning steps. Each step has a different list. Methyl Ethyl Ketone (MEK), solvent, B00148 is in all SOPM references for the cleaning of composite structures.

小任务 4　快速干燥修理零件有哪些方法？
Subtask 4　What are methods of quick drying procedure?

对可见的或吸收在构件内部的水分常有以下几种干燥方法。对于可见的水，直接擦拭、抽真空、压缩空气干燥。对于内部积水，常使用电热毯真空干燥、辐射热（烤灯）、热风干燥机、便携式烤箱或烘箱。每种方法都可去除水分，干燥方式的选择取决于修理条件。

Several drying methods are available for visible water and absorbed fluids. For visible water, mopping up, vacuuming, dry compressed air. For internal water, often use electric blanket vacuum drying, radiant heat (baking lamp), hot air dryer, portable oven or oven. Each of these methods can be applied, depending on repair conditions.

注意：实际航线维修针对干燥除湿有严格的工卡指令。目前，由于时间限制等因素，比赛技术文件并未严格要求进行真空干燥除湿，只需进行简单的擦拭或静置风干即可。本任务将详细描述手册从蜂窝夹芯层板、夹层板边缘带或平板结构中抽出水分的程序，具体比赛操作步骤是否包含此工序依据技术文件确定。

Note: Actual airline maintenance has strict time card instructions for drying and dehumidification. At present, due to time constraints and other factors, the technical documents of the competition do not strictly require vacuum drying dehumidification, only simple wipe or stand air drying can be. This training task will describe in detail the procedures for removing water from the honeycomb core sandwich panel, the sandwich panel edge band or a solid laminate panel in the manual. Whether this procedure is included in the specific operation steps of the competition will be determined according to the technical document.

材料与工具清单——清洁
Materials and Tools List-Cleaning-up

工具 / 材料 Tools/Materials	图示 Illustration
吸尘器 The vacuum cleaner 防尘、防毒面具 Dustproof gas mask 清洗剂（丙酮、三氯乙烷等） Cleaning agent (acetone, trichloroethane, etc.) 无绒布 Lint-free cloth 砂纸 Sandpaper 橡胶手套 Rubber gloves	

材料与工具清单——干燥
Materials and Tools List—Moisture Removal

工具 / 材料 Tools/Materials	图示 Illustration
热电偶 Thermocouples 压敏胶带 Pressure sensitive tape 电热毯 Heat blanket 金属网板 Metal perforated plate 透气毡 Breather cloth 两个真空接头 Two vacuum connections 真空袋 Vacuum bag 密封胶带 Sealant tape 剪刀 Scissor	

清洁任务工单
Cleaning-up Task Card

任务名称 Task Topic	清洁 Cleaning-up
步骤 Steps	规章 / 指令 Regulation/Instruction
1. 表面清洁 Surface Cleaning-up	若有油等污染，须清洁表面，将溶剂倒在擦拭纸或无绒干净的抹布上，使其浸润，然后擦拭待清洁的修理面。 If there is oil and other pollution, the surface needs to be cleaned, the solvent must be poured on the wiping paper or a clean cloth without lint, so that it can be soaked, and then wipe the repair surface to be cleaned. 擦拭清洁时需及时将溶剂擦干，不要让其在被擦拭表面自然挥发、干燥。重复上述的清洁操作，直到所有修理区域清洁为止。 The solvent should be wiped clean in time, do not let it be wiped on the surface of natural volatilization, dry. Repeat the above cleaning operation until all repaired areas are clean.
2. 再次清洁 Re-cleaning	若有粉尘等，使用吸尘器进行吸尘，条件允许再使用溶剂清洁。 If there is dust, use a vacuum cleaner for vacuuming, if conditions allow, then use solvent cleaning.
3. 检测清洗区域 Check Cleaning Area	修理区域的清洁质量可以通过无绒干净的抹布擦拭后状态大致判别，或者用水破裂试验确定。 The cleaning quality of the repaired area can be roughly determined by the status after wiping with a clean cloth without lint, or by the water rupture test.

真空除湿干燥任务工单
Vacuum dehumidification drying Task Card

任务名称 Task Topic	真空除湿干燥 Vacuum dehumidification drying
步骤 Steps	规章 / 指令 Regulation/Instruction
1. 准备 Preparation	切一块蜂窝芯材临时填满挖除区域。 Cut a piece of core to temporarily fill the damaged area for large repairs and put the core in the damaged area. 针对直径 4 in 或更小的损伤区域，使用最小 1/8 in 厚的带孔金属板，用以支撑真空袋。 针对直径 4 in (10 cm) 以上的损伤区域，用 0.16 in (4 mm) 或更厚的带孔金属板。 Use a perforated metal plate 1/8 in thick, minimum, to support the vacuum bag pressure for small damages 4 in or less wide diameter. Use a perforated metal plate that is 0.16 in. (4 mm) or thicker, for damage that is larger than 4 in. (10 cm). 注意：较薄的钢板会损坏修理部件和 / 或装袋设备。金属板必须是 3 in（8 cm），大于待修补区域。 Note: A thinner plate can damage the repair part and/or the bagging equipment. Metal plate must be 3 in. (8 cm) larger than the area to be repaired.
2. 放置热电偶 Thermocouple Installation	用压敏胶带固定热电偶。 Use pressure sensitive tape to hold the thermocouple(s) in position. 如果是单面修理，则至少在损坏区域的边缘放置两个热电偶。如果可以到达损伤面板的反面，则在未损坏的面板反面也需放置热电偶。 If one side is damaged, put at least two thermocouples at the edge of the damage on the damaged face-sheet. Put a thermocouple on the undamaged face-sheet opposite of the damage if both sides are accessible. 将热电偶放在电热毯的中心。 Put a thermocouple in the center of the heat blanket.
3. 真空封装 Vacuum bag assembly procedure	封装工装示意图参考图 3-1-3，各辅助材料尺寸参考图 3-1-4，各辅助材料放置顺序参考图 3-1-5。 Refer to Fig.3-1-3 for schematic diagram of Vacuum bag assembly. Refer to Fig.3-1-4 for dimensions，and refer to Fig.3-1-5 for laying-up sequence of each auxiliary material. 在热电偶上放置透气毡。透气毡必须至少比电热毯边缘大 2 in (5 cm)。 Apply breather material over the thermocouple(s). The breather material must be a minimum of 2 in. (5 cm) larger than the heat blanket(s) all around. 用压敏胶带固定透气毡。 Use pressure sensitive tape to hold the breather material in position. 将电热毯盖在透气毡上。 Put the heat blanket over the breather material. 在电热毯上方放置 4 或 5 层透气材料。透气毡必须至少比电热毯边缘大 2 in (5 cm)。 Apply 4 or 5 layers of breather material above the heat blanket(s). The breather material must be a minimum of 2 in. (5 cm) larger than the heat blanket all around. 在待修一侧，放置三个真空接头底座于透气毡之上，让待修区域夹在三个真空探头座之间。 On the side of the component to be repaired, put three vacuum probe bases on the surface breather with the repair area in between them. 再额外放置透气毡在真空接头底座、压力表和可调阀的下方。使用至少 4 层额外的透气毡。 Put additional breather material directly under the vacuum probe base, gage, and adjustable valve. Use a minimum of 4 layers of additional breather material.

任务名称 Task Topic	真空除湿干燥 Vacuum dehumidification drying
步骤 Steps	规章 / 指令 Regulation/Instruction
4. 密封真空袋 Seal the vacuum bagging	在修理区域周围贴上密封胶条。密封胶条与透气毡之间的距离约为 3 in (8 cm)。 Apply the vacuum sealant tape around the repair area. Put 3 in. (8 cm) distance between the vacuum bag sealant tape and the breather material. 制作一个大于待密封面积的真空袋。 Make a vacuum bag that is larger than the area to be sealed. 将真空袋材料放在修补区域的上方。 Put the vacuum bag material above the repair area. 用密封胶条密封真空袋的边缘。 Seal the edges of the vacuum bag with the vacuum sealing compound. 在真空袋材料上的真空接头和真空表将要安装的地方剪开一条缝，安装真空装置。 Cut slits in the vacuum bag material at the locations where the vacuum probes and the vacuum gage will be installed，install the vacuum hardware. 注意：狭缝要小，但需满足真空接头和真空表与其底座相连。 Note: Make the slits small, but sufficiently large to let the vacuum probes and gage attach to their bases.
5. 施加真空 Apply Vacuum	排除真空袋里的空气，调整阀门，保持大约 12 英寸汞柱 (305 毫米汞柱) 的真空度。 Evacuate the space under the vacuum bag(s) and adjust the valve to maintain a vacuum of approximately 12 inches of mercury (Hg)［305 millimeters of mercury (Hg)］.
6. 完成 Complete the Procedure	移除真空袋、电热毯、热电偶和透气毡。 Remove the vacuum bag, heat blanket, thermocouples, and breather cloth, from the repair area. 检查零件是否干燥。 Inspect the part to see if it is dry. （1）如果修复区域未干燥，更换蜂窝夹层组件的热电偶，然后再做步骤 3、步骤 4 和步骤 5。 If the repair area is not dry, replace the thermocouples, then do steps Step 3, Step 4 and Step 5 again. （2）如果零件足够干燥，则可继续进行后续维修工作。 If the part is sufficiently dry, then you can continue with the applicable repair.

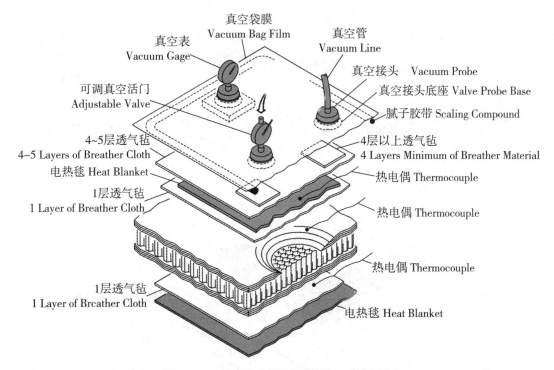

图 3-1-3　蜂窝夹芯结构除湿封装工装示意图

Fig.3-1-3　Schematic diagram of honeycomb sandwich structure dehumidification package

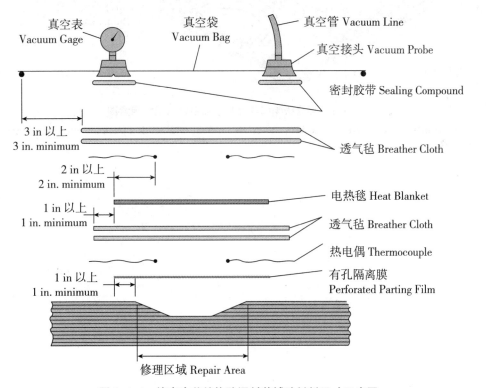

图 3-1-4　蜂窝夹芯结构除湿封装辅助材料尺寸示意图

Fig.3-1-4　Dimension diagram of auxiliary materials for honeycomb sandwich dehumidification package

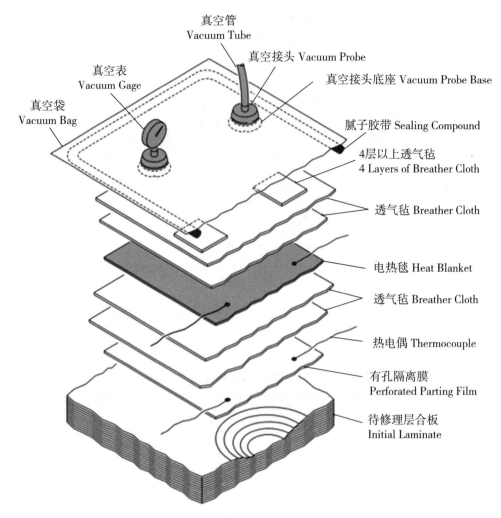

图 3-1-5　层合板结构除湿封装辅助材料尺寸示意图

Fig.3-1-5　Dimension diagram of auxiliary materials for dehumidification package of laminate structure

清洁与干燥练习题任务工卡

Cleanup and Absorbed Moisture Removal Exercise Task Card

（1）复合材料表面污染物的种类有哪些？如何清除？

What are the types of surface contaminants on composite materials? How can they be removed?

（2）复合材料在什么情况下需要进行清洁与干燥？

Under what circumstances do composite materials require surface treatment?

（3）复合材料表面清洁后，如何评估清洁效果是否满足要求？

How can the effectiveness of the cleaning process on the surface of composite materials be evaluated to ensure that it meets the requirements?

清洁与干燥操作工序卡

Cleanup and Absorbed Moisture Removal Operation Log

1. 工具、耗材、设备清点：

Inventory of tools, consumables, and equipment:

2. 清洁剂记录：

Cleaning agent record:

3. 清洁时间记录：

Record of cleaning time:

4. 除湿操作记录：

Drying operation record:

5. 除湿真空度、时间、温度记录：

Record of dehumidification vacuum level, time, and temperature:

6. 清点归还工具：

Inventory return of tools：

操作总结：

Summary：

子任务 2　去除损伤与打磨修理型面
Subtask 2　Damage Removal and Surface Preparation

【子任务解析 Subtask Analysis】

查阅零件制造图纸、表面清洁、干燥等工作完成后，就可以根据维修方案开始实施去除损伤工作，具体根据损伤报告中损伤深度，结合制造图纸判断损伤范围是否只在层合板内还是波及蜂窝芯，即需要实施去除层合板损伤，以及必要时去除蜂窝芯损伤。

Refer to parts manufacturing drawings, surface cleaning, drying after completion of work, can work according to maintenance plan started to remove damage, depending on the damage report damage depth combined with manufacturing drawings to determine whether a lesion is only within the laminate or spilled into the honeycomb core, namely the need to implement to remove laminate damage and if necessary, remove the damage to the honeycomb core.

通过本任务的训练，可掌握去除损伤的标准流程，具体包括去除漆层、层合板修理型面工序、蜂窝芯损伤等规范处理工序。

The standard process of damage removal can be mastered by this training, as well the removal of paint layer, laminate repair process, honeycomb core damage and other standard processing processes.

【子任务分解 Subtask Break-down】

（1）掌握去除漆层处理工序。

Master the process of removing paint layer.

（2）掌握去除蜂窝芯工序。

Master the process of removing honeycomb core.

（3）掌握层合板修理型面准备工序。

Master laminate to lay layer damage process.

85

小任务 1 为什么需要去除漆层以及注意点？

Subtask 1 Why should the paint layer be removed and what should be noted?

首先去除漆层可以让损伤目标更加凸显，维修区域与非维修区域做区隔；其次，漆层的存在不利于胶接质量和封装的气密性等。

Firstly, Removing the paint layer can make the damage target more prominent, and the maintenance area is separated from the non-maintenance area; Secondly, the existence of paint layer is not conducive to bonding quality and encapsulation of air tightness.

需要注意的是，去除漆层时通常需要根据损伤的大小划定去漆层区域，一般漆层区域会比最大修理区域边缘超出 1 in（25.4 mm）。注：具体数值应严格遵守 SRM 文件或者指定的参考维修资料。去除漆层一般使用砂纸打磨，此过程注意粉尘防护，操作人员需要佩戴防尘口罩与防飞溅的护目镜。

One thing to note: The paint layer removal area usually needs to be defined according to the size of the damage. Generally, the paint layer area will be 1 in. (25.4 mm) larger than the edge of the maximum repair area. Note: The specific value should be strictly in accordance with the SRM file or specified reference maintenance data. The removal of paint layer usually uses sandpaper grinding, this process pays attention to the dust method, the operator needs to wear dust mask and splash goggles.

零件保护是完成个人防护的后一步，即在实施打磨前确保零件其他正常区域得到保护，例如，去漆区域以外用胶带保护，妨碍打磨工序的紧固件、装配件预先拆除，相应的装配孔进行封胶保护等。

Parts protection, that is, ensure that other normal areas of the parts are protected before the implementation of grinding, such as tape protection outside the paint removal area, pre-removal of fasteners and assemblies that hinder the grinding process, protection of the corresponding assembly holes, etc.

小任务 2　损伤的形状各式各样，但是去除打磨形状为何一般为圆形与跑道圆形式？

Subtask 2　The shape of the damage varies, but what is the reason that the removal of the grinding shape is generally circular and racetrack round form?

首先，SRM 规定要求去除损伤、修理区域以及修理层等加工成圆形与长圆形式，这是因为圆形或者长圆形没有棱角不容易出现应力集中；其次，去除过程可使用气动旋转打磨，圆形与长圆形更容易加工。

Firstly, The SRM will require the removal of damage, repair areas and repair layers to be processed into round and oblong forms, which are round or oblong without edges and corners and are not prone to stress concentration; Secondly, the removal process can use pneumatic rotary grinding, circular and oblong easier to process.

小任务 3　如何使用去除蜂窝芯损伤切割工具镂铣机？使用时需要注意什么？

Subtask 3　How to use a cutting tool to remove honeycomb core damage? What need to be careful when using?

镂铣机又称气动铣，常用来驱动铣头、铣刀等工具，适用于复合材料切割、修边和损伤部位去除等修理工作，大多数镂铣机以 20 000~30 000 r/min 为有效转速，其工装示意图如图 3-2-1 所示。

Router, also known as pneumatic router, is commonly used to drive milling head, milling cutter and other tools, commonly used composite material cutting, trimming and the removal of damage parts and other repair work, most of the engraving and milling machine, and the schematic diagram of its tooling is shown in Fig.3-2-1.

针对镂铣深度，SRM 上建议如果整个蜂窝芯都需移除，并且假设蜂窝芯以下的层合板没有损坏，那么当蜂窝芯挖除剩下 0.02~0.04 in（0.5~1.0 mm）时，镂铣操作应该停止，以防止对下层面板造成损伤。剩余的蜂窝芯应小心地进行手动打磨，直至黏合层（胶膜）。只要胶膜完好无损，就可保留不必打磨。为实现较好的修理效果，可轻轻打磨胶膜上胶瘤，获得平整的蜂窝安装面。

Regarding the cutting depth, the SRM recommends that if the entire honeycomb core needs to be removed and assuming that the opposite face is not damaged, then the cutting operation should stop when approximately 0.02 −0.04 in. (0.5−1.0 mm) of the core is remaining. This will prevent the opposite skin from being damaged. The remaining core should be carefully removed by hand sanding down to the adhesive layer. The adhesive layer can be left providing it is in good condition and undamaged. Lightly abrade the remaining adhesive layer to remove adhesive fillets and to obtain a flat surface.

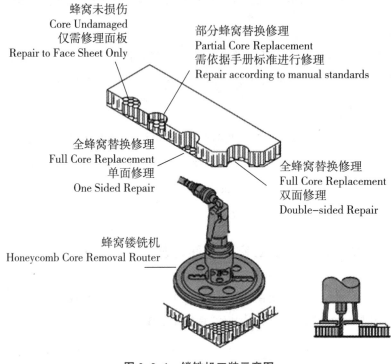

图 3-2-1　镂铣机工装示意图

Fig.3-2-1　Router illustration

小任务 4　打磨安全注意事项有哪些?
Subtask 4　What are the safety precautions for sanding procedure?

（1）呼吸系统防护。打磨的粉尘对人体的呼吸系统有害，在打磨时佩戴呼吸保护用品，如防尘口罩、护目镜和耳塞。打磨要在吸尘间进行，或者吸尘与打磨同时进行。

Respiratory system protection. Sanding dust is harmful to the human respiratory system. Wear respiratory protection such as dust masks, goggles and earplugs while sanding. Polishing should be carried out in the vacuuming room, or vacuuming and polishing at the same time.

（2）皮肤保护。细小的纤维丝和打磨粉尘进入皮肤会引起发炎和瘙痒，打磨时要注意保护裸露的皮肤，需穿全身防护服。

Skin protection. Fine filaments and grinding dust entering the skin can cause inflammation and itching. Be careful to protect exposed skin when polishing. Wear full body protective clothing.

（3）眼睛防护。戴安全防护眼镜可以保护脸前方和侧面来的冲击伤害，如飞溅的液体和灰尘或者热辐射的损伤。如果眼部被液体或者灰尘损害，马上用大量生理盐水或干净水冲洗，车间都要求在显眼位置配置洗眼设备，最简单的是瓶装的洗眼水。

Eye protection. Wear safety glasses to protect the front and side of the face from impact damage, such as splashing liquid and dust, or from thermal radiation. If the eyes are damaged by liquid or dust, wash them immediately with a large amount of normal saline or clean water. The workshop requires that eye washing equipment be configured in a conspicuous location, the simplest is bottled eye washing water.

材料与工具清单——清除损伤
Materials and Tools List-damage removal

工具 / 材料 Tools/Materials	图示 Illustration
镂铣机 Router 气动打磨机 Pneumatic grinding machine 真空吸尘器 The vacuum cleaner 剪刀 Scissor 防护服 Protective Clothing 防尘、防毒面具 Dustproof gas mask 清洗剂（丙酮、三氯乙烷等）Cleaning agent (acetone, trichloroethane, etc.) 无绒布 Lint-free cloth 砂片 Sandpaper 橡胶手套 Rubber gloves 压敏胶带 Pressure sensitive tape	

湿法芯材填充

去除面板和蜂窝芯损伤任务工单
Remove the Damaged Skin and Core Task Card

任务名称 Task Topic	去除面板和蜂窝芯损伤 Remove the Damaged Skin and Core
步骤 Steps	规章 / 指令 Regulation/Instruction
1. 去除受损蜂窝芯材 Remove the damaged core	蜂窝去除大小与面板切口大小保持一致。 Remove the core to the same dimension as the cutout in the skin. 设置镂铣机深度到必要的值，以去除损坏的蒙皮和芯材。 Set the depth of the router cutter to the necessary value to remove the damaged skin and core. 注意：使用镂铣机时，不要使切割处粘接区域过热。 Caution: Do not cause the bond area to get overheat when you use the router. 不要对另一侧蒙皮造成初始损伤 (或额外损伤)。 Do not cause initial damage (or additional damage) to the opposite skin. 确保将所有的损坏都清除掉。 Make sure that all of the damage is removed. 全深度芯损的清除修整，需轻磨芯材所在处底部的黏合面 (胶膜)。 As for full depth core damage removal trimming, lightly abrade the adhesive surface at the bottom of the hole (where the core was). 注意：除非不满足修理效果，否则不需要将底部的胶粘剂全部去除。如果胶粘剂受到水或其他液体的损坏或蒙皮缺损，则粘接效果将认定为不能满足性能要求。 Note: It is not necessary to remove all of the adhesive from the opposite face-sheet unless it is in an unsatisfactory condition. The adhesive is unsatisfactory if it has damage from water or other fluids, or if bare skin shows.
2. 清洁表面 Cleaning-up	使用真空吸尘器将蜂窝板内的灰尘和颗粒从胶粘剂（胶膜）表面清除。先用吸尘器清除损坏区域的所有颗粒、灰尘、油脂、油、液体、湿气或其他污染，然后用无绒的干净布擦拭。 Use a vacuum cleaning device to remove dust and particles from the adhesive surface. Remove all of the particles, loose dirt, grease, oil, fluids, moisture or other contamination in the damaged area with a vacuum cleaner first and then a lint-free clean cloth.
3. 修理型面设计 Sanding plan installation	找出将缺陷包围起来的最小形状，可参考图 3-2-2。 Find the smallest shape that will enclose the defect, refer to Fig.3-2-2. 在零件上标记打磨形状，圆角半径最小为 0.5 in (13 mm)。 Mark the shape on the part, The minimum-fillet radius is 0.5 in. (13 mm). 确定要移除的面积大小，并记录打磨参数。 Determine size of area to be removed and record the sanding parameters. 在修理区域边缘外大约 1 in (25 mm) 处用压敏胶带粘贴保护。 Apply pressure sensitive tape to mask off the area approximately 1 in. (25 mm) from the edge of where you will remove the damage.
4. 去漆层并 制备修理型面 Remove the paint and sanding in the area to be inspected	从修理区域除去油漆和底漆、涂层、箔和密封剂（如果适用），至少为修理区域边缘外各方向扩展至少 2 in (5 cm)。 Remove the paint and primer, coatings, foil and sealant (if applicable) from the defective area, plus at least 2 in. (5 cm) in all directions. 使用 150 号砂纸或更细的氧化铝砂纸进行打磨，直到看到漆层下面的表面层，但不要用砂纸打磨掉表面层。 Use No.150 grit or finer aluminum oxide abrasive paper. Sand until the surfacer layer comes into view below paint but do not sand the surfacer. 注意：不要对修理区域以外的结构层造成损坏。 Note: Be careful not to cause damage to the structural plies outside of the area to be removed.

任务名称 Task Topic	去除面板和蜂窝芯损伤 Remove the Damaged Skin and Core
步骤 Steps	规章 / 指令 Regulation/Instruction
5. 完成 Completion	完成此工序后，要及时关闭设备，清理工具和航材。 To timely close equipment after completion of the process, cleaning tools and materials.

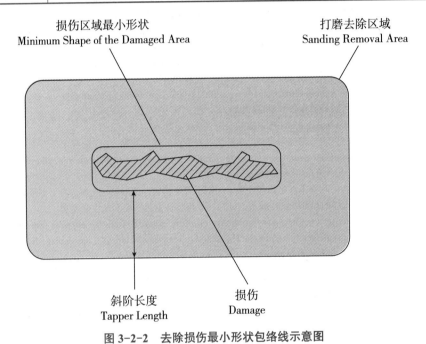

图 3-2-2　去除损伤最小形状包络线示意图

Fig.3-2-2　Scheme for minimum enclosure area for damage removal

修理型面准备任务工单（永久修理）
Surface Preparation for Permanent Repair Task Card

任务名称 Task Topic	修理型面准备—单面蒙皮 / 单面蒙皮＋蜂窝芯材 Surface Preparation for Permanent Repair—Damage to Skin only / one Skin+Honeycomb Core
步骤 Steps	规章 / 指令 Regulation/Instruction
1. 描绘损伤轮廓并去除损伤 Outline the damaged area and remove	针对单面蒙皮损伤：描出需要去除的损伤区域。 For skin (surface only): Outline the damaged area to be removed. 针对蒙皮和蜂窝芯材损伤：按照每层材料相应的步骤勾画出必须去除的区域，可参考图 3-2-3 阶梯打磨——单面蒙皮损伤圆形修理和图 3-2-4 斜阶打磨——单面蒙皮损伤圆形修理图示，但一般原则适用于任何修复形状。 For skin and honeycomb: Outline in steps corresponding to each ply of material that must be removed. Ref. Fig. 3-2-3/3-2-4 Step/Scarf Preparation for Bonded Repair–Skin Damage on one Face show a circular repair but the general principles are applicable to any repair shape. 注意：对于非圆形的修复，请确保角部有 0.5 in（13 mm）的半径。不允许有方形的角。对于小损伤修复，您可以使用较小的半径，或者将其视为圆形损伤处理。 Note: For non-circular repairs make sure that corners have a radius of 0.5 in. (13 mm). Square corners are not permitted. For small damage repairs you can use reduced radius, or you can treat it as a circular damage.
2. 精细修整 Fine trimming	用吸尘器清除所有的杂质。 Remove all waste material with a vacuum cleaner. 从 280 目到 400 目，用（耐水碳化硅）磨料布精细打磨修补部位。 Abrade the repair area with water resistant silicon carbide abrasive cloth. Start with 280 grades and finish with 400 grades.
3. 检查损伤完全去除 Verify the damage has been completely removed.	在初始损伤被去除后，再次检查被去除区域的平整部分，确保所有损伤已经被清除。 After the initial damage has been removed, check the flat section of the removed area again to ensure that all damage has been removed. 如果你能看到水、油、燃料或其他液体，必须清除。 If you can see water, oil, fuel or other liquids, you must remove them.
4. 去除胶带保护并干燥 Unmask and Dry	去除胶带保护并清洁打磨斜面。 Unmask and clean the scarfed surface.
5. 修理表面保护 Surface protection	零件干燥。 Dry the structure. 保护修理区域的表面，防止表面污染。 Protect the surface in the repair area to prevent contamination.
6. 完成 Completion	完成此工序后，要及时关闭设备，清理工具和航材。 To timely close after completion of the process equipment, cleaning tools and materials.

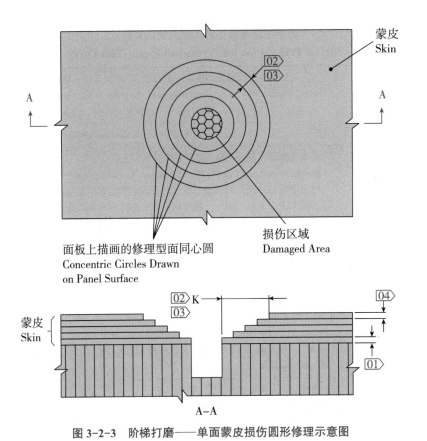

图 3-2-3　阶梯打磨——单面蒙皮损伤圆形修理示意图

Fig.3-2-3　Step Preparation for Bonded Repair—Skin Damage on one Face Show a Circular Repair

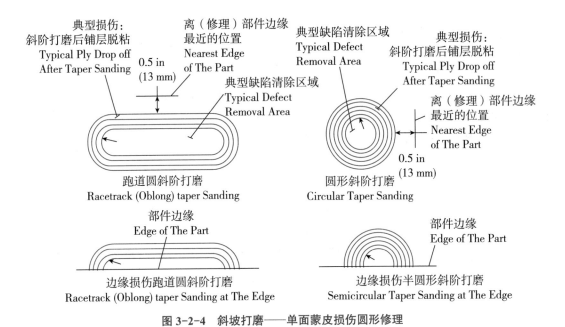

图 3-2-4　斜坡打磨——单面蒙皮损伤圆形修理

Fig.3-2-4　Scarf Preparation for Bonded Repair—Skin Damage on one Face Show a Circular Repair

注释：

（1）"尺寸 K" 不包括最终的表面修补层尺寸。

"Dimension K" does not include the final surface repair plies.

（2）对于织物结构，打磨后应保留第一层铺层（紧贴蜂窝的铺层）。对于单向带结构，打磨后应保留一到两层铺层，并与最后一层铺层保持 0.1~0.3 mm 的台阶高度（一般默认单向带的厚度是织物的一半，实际生产中常取 0.125 mm，而织物的厚度常取 0.25 mm）。

For fabric construction, the first ply should remain after step sanding. For tape construction, one or two plies should remain after step sanding to provide a 0.1−0.3 mm step（In general, we default that the thickness of the unidirectional tape is half of the fabric, and the in practicable manufacture often takes 0.125mm, while the thickness of the fabric is often 0.25mm）.

（3）"K" 的尺寸在特定章节中给出。

The dimensions of "K" are given in the specific chapter.

（4）每个修复层都要进行 13 mm 宽度的台阶打磨，并遵守步骤 1 的要求。

Step sand 13 mm for each repair ply and observe the requirements of step 1.

（5）镂铣机的镂铣深度应与修复层的厚度相匹配。

The depth of the cutter should match the thickness of the repair ply.

去除损伤与打磨修理型面练习题任务工卡
Damage Removal & Repair Surface Preparation ExerciseTask Card

（1）复合材料去除损伤和打磨修理型面时需要注意哪些安全措施？

What safety measures should be taken when removing damage and sanding the repair area on aircraft composite materials?

（2）复合材料去除损伤和打磨修理型面时如何保持修理区域的平整度？

How can the flatness of the repair area be maintained when removing damage and sanding the repair area on aircraft composite materials?

（3）飞机复合材料去除损伤和打磨修理型面时如何控制修理区域的尺寸和深度？

How can the size and depth of the repair area be controlled when removing damage and sanding the repair area on aircraft composite materials?

去除损伤与修理型面准备操作工序卡

Removal of Damage and Repair Surface Preparation Operation Log

1. 工具、耗材、设备清点：

Inventory of tools, consumables, and equipment:

2. 阶梯打磨尺寸设计：

Design of step polishing dimensions:

3. 阶梯打磨尺寸控制：

Control of step polishing dimensions:

4. 去除蜂窝尺寸设计或 N/A：

Design of cell removal dimensions or N/A:

5. 去除蜂窝尺寸控制记录或 N/A：

Record of cell removal dimension control or N/A:

6. 蜂窝芯材方向标记或 N/A：

Marking of cell core material direction or N/A:

7. 清点归还工具：

Inventory return of tools:

操作总结：

Summary:

任务 3 去除损伤考核表

姓名		班级		
评价维度	分值/分	自评（30%）	互评（30%）	师评（40%）
素养（20%）				
1. 材料清点齐全	5			
2. 任务书自学情况	5			
3. 安全文明操作及 6S	10			
技能（60%）				
1. 真空干燥除湿操作规范性（封装顺序、辅助材料尺寸、封装真空度）	30			
2. 去铺层设计正确性（数量、尺寸设计）	15			
3. 去铺层打磨平整度（第一层与第二层厚度差、各层尺寸容差 4%）	15			
总结报告（20%）				
1. 训练总结的完整性	10			
2. 个人反思与拟订后续训练计划	10			
总计	100			
任务完成情况	提前完成			
	准时完成			
	滞后完成			

损伤修补
Damage Repair

【任务情境 Task Scenario 】

你是 WS 航空公司的航线结构修理部门操作员，并持有复合材料结构件修理授权书。某日你到岗后，收到航线报机身右侧壁板存在凹坑损伤，深度超过损伤容限，已完成阶梯打磨、完全去除损伤工序。你的工作是按照手册要求完成维修。

You are an operator in the line structural repair department of WS Airlines and hold a repair authorization for composite structural components. One day, after you arrive at work, you receive a report of a dent on the right-hand side wall panel of the aircraft, with a depth that exceeds the damage tolerance. The step grinding process has been completed to remove the damage completely. Your task is to complete the repair according to the manual requirements.

【任务解析 Task Analysis 】

损伤修补是进行复合材料结构损伤维修工作的第四步。

Damage repair is the fourth step in performing composite structure repair work.

损伤修补工作需要根据先前的探伤检测结果和修理方案进行具体操作。常见的损伤修补方法包括补片、填充、热固性树脂注入等。在选择具体的修补方法时，需要考虑到损伤的类型、大小、位置，以及修补后结构强度和气动外形要求。

The damage repair work needs to be carried out based on the previous inspection results and repair plan. Common damage repair methods include patching, filling, injection of thermosetting resin, etc. When choosing a specific repair method, it is necessary to consider the type, size, and location of the damage, as well as the structural strength and aerodynamic shape requirements after the repair.

完成本任务训练后，应该能够实现以下目标：

After completing training for this task, the following goals should be achieved:

知识目标 Knowledge Objectives

（1）知道湿铺层的定义与铺叠方式。

Understand the definition and laying methods of wet layup.

（2）知道注射修理、预浸料修理的流程。

Understand the process of injection repair and prepreg repair.

（3）熟悉影响修理效果的各项指标。

Familiar with various indicators that affect repair effectiveness.

能力目标 Ability Objectives

（1）掌握湿法贴补修理的标准施工流程，掌握含胶量的计算方法。

Master the standard construction process of wet patch repair and the calculation method of adhesive content.

（2）按照相应的结构修理程序，正确地进行蜂窝安装。

Correctly install the honeycomb according to the corresponding structural repair procedure.

（3）能进行损伤修补后的检验和质量控制，并保证修补后的结构符合要求。

Able to carry out inspection and quality control after damage repair, and ensure that the repaired structure meets the requirements.

素质目标 Emotion Objectives

（1）具有细心和耐心的品质，复合材料制造过程中需要对每个工艺环节进行精确控制，以保证制品精度。

Carefulness and patience are required in the manufacturing process of composite materials, and each process must be precisely controlled to ensure the accuracy of the product.

（2）团队合作与沟通能力：学会与团队成员配合，共同完成修理任务。培养与客户、同事、监察有效沟通的能力。

Team cooperation and communication skills: learn to cooperate with team members to complete maintenance tasks. Develop the ability to communicate effectively with customers, colleagues, and supervisors.

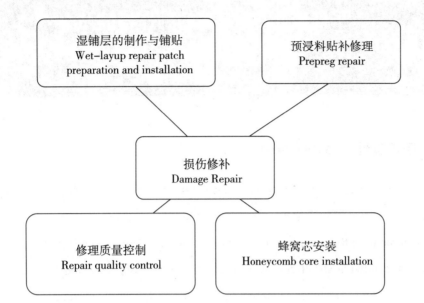

湿铺层的制作与铺贴
Wet-layup repair patch
preparation and installation

预浸料贴补修理
Prepreg repair

损伤修补
Damage Repair

修理质量控制
Repair quality control

蜂窝芯安装
Honeycomb core installation

子任务 1 湿铺层的制作与铺贴
Subtask 1 Wet-layup Repair Patch Preparation and Installation

【子任务解析 Subtask Analysis】

湿法贴补修理是航线复合材料层合板结构以及蜂窝夹层结构维修最常见的修理方法。湿铺层指纤维干布进行树脂浸渍后，与修理型面尺寸贴合的修理层补片。

Wet patch repair is the most common repair method for route composite laminate structures and honeycomb sandwich structures. Wet patching refers to impregnating fiber dry cloth with resin and applying the repair layer patch that fits the size of the repair surface.

通过本任务的训练，可以掌握湿铺层制备的标准流程，具体包括胶液配制、浸渍纤维制作湿铺层以及湿铺层的铺贴。

Through training in this task, we can master the standard process of preparing wet patches, including adhesive preparation, impregnated fiber preparation for wet patches, and wet patch installation.

【子任务分解 Subtask Break-down】

（1）理解胶液配制方法与规范。

Understand adhesive preparation methods and specifications.

（2）掌握带铺层设计的湿铺层制备方法。

Master the wet patch preparation method with ply design.

（3）掌握带铺层设计的湿铺层铺贴方法。

Master the wet patch installation method with ply design.

小任务 1 什么是湿铺层？
Subtask 1 What is the meaning of wet-laying-up plies?

湿铺层指湿法修理工艺制作的修理补片，是树脂浸渍干布纤维形成的一种未固化的补片。

Wet-laying-up plies, is an uncured repair patch made by impregnating dry cloth fibers with resin.

小任务 2 修理常用的胶粘剂种类
Subtask 2 Types of commonly used adhesives for repair

层压树脂为双组分、低黏度的环氧树脂，主要用来浸渍干织物和注入凹坑或分层修补。

Laminating resins are two component, low viscosity, epoxy resins. They are used to impregnate dry fabric and to inject dent or delamination repair.

胶膜是环氧基材料，用于预浸料共粘接，或用于复合材料或金属材料的粘接。

Adhesive films are epoxy based materials, used for co-bonding of prepregs, or for bonding composite or metallic materials.

小任务 3 蜂窝修复工作，常用到哪些原料？
Subtask 3 What materials are commonly used for honeycomb repair work?

（1）胶粘剂 adhesives

（2）层合板用树脂 Laminating resins

（3）低黏度胶粘剂 Low viscosity adhesives

（4）低密度化合物 Low density compounds

（5）干织物 Dry Fabric

（6）预浸料 Prepreg

（7）发泡胶 Foaming adhesives

（8）胶膜 Adhesive films

（9）防腐底漆 Anti-corrosion Primers

（10）预固化补片和型材 Pre-cured doublers and sections

（11）蜂窝芯材 Honeycomb core

复合材料层合板的材料性能和典型数据是通过测试或使用所提供的材料数据库获得的。层合板的性能可以从代表尽可能接近贴片修补中使用的层合板的测试结果中得到。所有属性都应使用特征材料值。

Material properties and typical data for composite laminates are obtained by testing or by using the material database provided. The laminate properties can be obtained from test results of laminates that represent the laminate used in the bonded patch repair as closely as possible. Characteristic material values should be used for all properties.

（1）为了表征层合板的特性，需要提供以下信息：

To characterize the properties of the laminate, the following information should be provided:

① 纤维类型 Fiber type

② 织物类型 Type of fabric

③ 树脂类型（如环氧树脂、聚酯树脂、聚氨酯等）

Resin type (such as, epoxy, polyester, polyurethane, etc.)

④ 物料存放条件：温度、层合板含水率（湿、干）

Material storage condition: temperature, moisture content of the laminate (wet, dry)

⑤ 制备工艺：方法、温度、压力、真空、后固化（温度、时间）

Preparation process: method, temperature, pressure, vacuum, post curing (temperature and time)

⑥ 纤维取向、铺层方式、体积分数的控制

Control of fiber orientation, layer sequence and volume fraction

⑦ 孔隙率 Void ratio

用于结构修补的胶粘剂应按照制造商的建议使用。所有胶粘剂的详细信息包括处理、混合和应用胶粘剂，应在材料数据表和修复计划中详细说明。

Adhesives used for structural repairs should be in accordance with the manufacturer's recommendations. The details of all adhesives, including the handling, mixing, and application of adhesives, should be specified on the Material Data Sheet and on the repair plans.

（2）为了表征胶粘剂的特性，应提供以下信息：

To characterize the adhesive properties, the following information should be provided:

① 胶粘剂材料成分 Constituent adhesive material

② 胶粘剂类型（如环氧树脂）Adhesive type (e.g., epoxy)

③ 具体胶粘剂类型 Specific adhesive type

④ 催化剂（商品名称、批号）Catalyst (trade name and batch number)

⑤ 填料（商品名称和批号）Fillers (trade name and batch number)

⑥ 添加剂（商品名称、批号）Additives (trade name and batch number)

⑦ 加工方式：方法、温度、压力 Processing: method, temperature, pressure

⑧ 固化温度和时间 Curing temperature and time

⑨ 后固化（温度和时间）Post curing (temperature and time)

⑩ 芯材密度 Density of core material

⑪ 玻璃化转变温度 Glass transition temperature

小任务 4　湿铺层修理质量控制的具体方法有什么？

Subtask 4　What are the specific methods for quality control of wet layup repairs?

（1）确认修理面积：在进行湿铺层修理前，需要对损伤部位进行评估和测量，确认需要修理的面积，并进行标记。

Confirm repair area: Before wet laying-up, it is necessary to evaluate and measure the damaged area, confirm the area that needs repair, and mark it.

（2）预处理：在进行湿铺层修理前，需要对损伤部位进行预处理，包括去除污垢、油脂等杂质，利用砂纸或其他工具将损伤部位打磨平整，并用洁净的布清洁干净。

Pre-treatment: Before wet laying-up, the damaged area needs to be pre-treated, including removing dirt, grease, and other impurities, sanding the damaged area smooth with sandpaper or other tools, and cleaning it with a clean cloth.

（3）确保环境条件稳定：湿铺层的制备需要一定的湿度和温度条件，要确保环境中的湿度和温度稳定，避免影响铺层的质量。

Ensure stable environmental conditions: The preparation of the wet laying-up requires certain humidity and temperature conditions. Make sure that the humidity and temperature in the environment are stable to avoid affecting the quality of the layer.

（4）准备湿铺层材料：根据修理方案准备好所需的湿铺层材料，包括预浸料、织物、树脂等，确保其符合要求，并在使用前按照要求进行干燥、搅拌等处理。例如，精确测量树脂和固化剂的比例。树脂和固化剂的比例对于复合材料湿铺层的质量至关重要，需使用准确的称重仪器和比例计，精确测量出二者的比例。

Prepare wet lay-up layer materials: Prepare the required wet lay-up layer materials according to the repair plan, including pre-impregnated materials, fabrics, resins, etc., ensuring that they meet the requirements, and drying, stirring, etc. before use according to the requirements. For example, accurately measure the ratio of resin to curing agent. The ratio of resin to curing agent is crucial for the quality of the wet lay-up layer of composite materials. Accurate weighing instruments and proportion calculators should be used to accurately measure the ratio of the two.

（5）铺层操作：根据修理方案进行湿铺层操作，确保每一层的铺层方向、层数、面积等符合要求，同时注意控制铺层的厚度和平整度。湿铺层的厚度需要控制在一定的范围内，过厚或过薄都会影响铺层的性能。通过在模具上设置厚度控制器或使用测厚仪等工具，可以精细控制铺层的厚度。纤维布的平整性对于湿铺层的质量同样重要。在铺设纤维布料时，需要将其平整展开，避免出现褶皱或气泡。

Laying-up operation: Conduct the wet lay-up layer operation according to the repair plan, ensuring that the direction, number of layers, and area of each layer meet the requirements, while also controlling the thickness and smoothness of the layer. The thickness of the wet lay-up layer needs to be controlled within a certain range. Being too thick or too thin can affect the performance of the layer. By setting a thickness controller on the mold or using tools such as thickness gauges, the thickness of the layer can be precisely controlled. The smoothness of the fiber cloth is also important for the quality of the wet lay-up layer. When laying the fiber cloth, it needs to be laid flat to avoid wrinkles or bubbles.

（6）固化和后处理：在湿铺层完成后，需进行固化和后处理。对于预浸料修复，需要按照要求进行升温固化，且铺层固化的时间和温度都会影响其性能。通过控制固化时间和温度，可以确保铺层固化充分，同时避免出现二次固化。固化完成后，需要进行打磨、喷漆等后处理操作，使修复后的结构表面平整、外观无缺陷。

Curing and post-treatment: After the wet lay-up layer is completed, it needs to be cured and post-treated. For pre-impregnated repairs, it needs to be cured by heating according to the requirements, and the time and temperature of layer curing will affect its performance. By controlling the curing time and temperature, it can ensure that the layer is fully cured and avoid secondary curing. After curing, post-treatment operations such as sanding and painting are needed to make the repaired structure surface flat and free of defects.

（7）质量检测：在湿铺层修理完成后，需要对修复部位进行检验，包括目视检查、尺寸测量、物理性能测试等，确保修复后的结构符合要求，并记录检验结果和修复情况，以备后续参考和分析。

Quality inspection: After the wet lay-up layer repair is completed, the repair area needs to be inspected, including visual inspection, dimensional measurement, physical performance testing, etc., to ensure that the repaired structure meets the requirements, and record the inspection results and repair situation for future reference and analysis.

（8）记录和追溯控制：需要对整个修复过程进行详细记录，并建立修复记录和追溯体系，以确保质量的可追溯性和修复过程的可控性。

Recording and traceability control: The entire repair process needs to be recorded in detail, and a repair record and traceability system should be established to ensure the traceability of quality and the controllability of the repair process.

材料与工具清单——湿铺层制备与铺贴

Materials and Tools List—Wet-layup Repair Patch Preparation and Installation

材料 / 工具 Materials/Tools	图示 Illustration
剪刀 Scissor 玻璃纤维 / 碳纤维 GF/CF 树脂基体 Resin 电子秤 Electronic scale 刮板 Scraper 量杯 Measuring cup 搅拌棒 Stirring rod 防护服 Protective Clothing 清洗剂（丙酮、三氯乙烷等） Cleaning agent (acetone, trichloroethane, etc.) 无绒布 Lint-free cloth 砂片 Sandpaper 橡胶手套 Rubber gloves 压敏胶带 Pressure sensitive tape	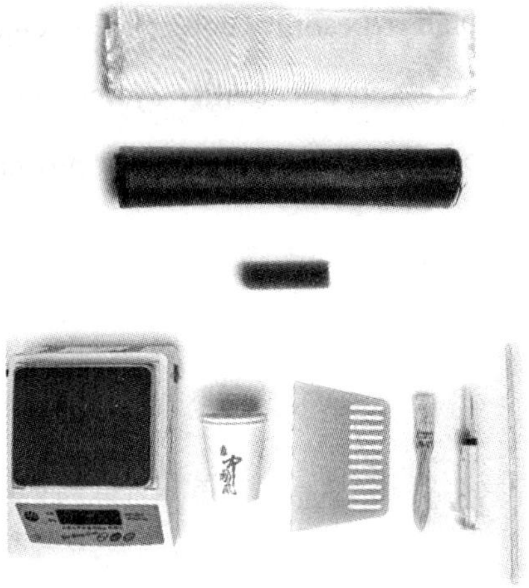

湿铺层制备任务工单
Wet-layup Repair Patch Preparation Task Card

任务名称 Task Topic	湿铺层制备 Wet-layup repair patch preparation
步骤 Steps	规章 / 指令 Regulation/Instruction
1. 胶液配制 Adhesive preparation	在混合树脂 / 胶粘剂之前，必须先完成准备工作。 Preparation work must be completed before mixing resins/adhesives. 计算完成维修所需的树脂量。确保树脂和固化剂混合后能在施工时间 80％ 以内使用完。 Calculate the quantity that is necessary to complete the repair. Ensure that the quantity mixed can be used before 80 % of the specified pot life has elapsed. 在打开容器之前，确认材料已达到室温。在取出所需的树脂 / 胶粘剂和固化剂后，立即重新安装容器盖。 Confirm that the material has reached ambient temperature before opening the container. Ensure that container lids are refitted immediately after removing the required amount of resin/adhesive and hardener. 准确称量树脂 / 胶粘剂和固化剂，误差范围在 ±2% 之内。需使用干净、无蜡无油的容器进行混合。 Weigh the resin/adhesive and hardener within an accuracy of ±2%. Use clean, wax and oil free containers for mixing. 为了确保最佳性能，需混合均匀树脂 / 胶粘剂和固化剂，至少混合 5 min，以得到均匀的混合物。确保容器壁上的胶粘剂也得到充分混合。混合过程中避免产生气泡。如有必要，使用真空技术去除气泡。 To ensure optimum properties, mix the resin/adhesive and hardener together for a minimum of 5 minutes to get a uniform homogenous mixture. Ensure that the material against the side of the container is included in the mixture. Avoid entrapment of air bubbles during mixing. Use vacuum techniques to degas if necessary. 注意：如有需要，在混合胶粘剂时可加入 2%~5% 的增稠剂，以获得不流动的糊状物。但这不适用于低密度树脂。 Note: If required, include 2% to 5 % by weight of thickening agent to the mixed adhesive while mixing, to give a paste that does not flow. Not applicable to low density compounds.
2. 铺层准备 Laminating preparation	对于碳纤维织物，其与胶液的质量比为 1 ∶ 1.3。对于玻璃纤维织物，其与胶液的质量比为 1 ∶ 1。实际称量时需考虑冗余设计或应对意外情况，将所需胶液质量增加 30%，以获得 CFRP 50%±5% 或 GFRP 42%±5% 的质量含量。 Give a fabric to mixed material ratio of 1:1.3 by weight for carbon fabric and 1:1 for structural glass fabric. Consider that the required amount of mixed material is increased by 30% for contingency purposes. The goal is to obtain a rein content by weight of 50% ± 5% with CFRP and 42% ± 5% with structural GFRP. 混合少量树脂 / 胶粘剂和固化剂容易产生误差，除非使用高精度称重设备。通常胶液至少为 50 g（2.0 盎司），100 g 为最佳（4.0 盎司），则可以将产生误差的概率降至最低。对于非常小的修复工作，可能会导致浪费。 Mixing small quantities of resin/adhesive and hardener carries the risk of errors unless accurate weighing devices are used. The risk is minimized if at least 50 g (2.0 oz.), but preferably 100 g (4.0 oz.) of resin/adhesive and the corresponding amount of hardener is mixed. On very small repairs this may lead to some wastage.
3. 湿铺层裁剪 Wet laying-up repair patch preparation	（1）参考维修方案，确定所需修补层数、材料性质和铺层方向。 Refer to repair scheme to determine the number, style and orientation of the required repair plies.

任务名称 Task Topic	湿铺层制备 Wet-layup repair patch preparation
步骤 Steps	规章 / 指令 Regulation/Instruction
3. 湿铺层裁剪 Wet laying-up repair patch preparation	（2）裁剪足够大的纤维干布，以满足已确定的修理面积需求。 Cut a large enough piece of dry fabric to meet the requirements established for the fabric selection. （3）裁剪两张隔离膜，尺寸比纤维干布大 150 mm 左右（6 in）。 Cut two pieces of parting film approximately 150 mm (6 in.) larger in dimension than the cut fabric. （4）在平整的桌面上贴上一隔离膜。 Tape one piece of the parting film on a smooth flat surface. 注意：确保隔离膜没有褶皱。 Note: Make sure that the parting film is free from wrinkles. （5）准备适量的胶液。 Prepare necessary quantity of laminating resin. （6）将混合好的胶液涂抹在隔离膜上，使用刮板均匀涂开树脂。 Pour the resin mix on the parting film and distribute it uniformly using a squeegee. 将 80%±5% 的胶液涂在隔离膜上，并保留少量树脂，在维修区域涂覆一层薄薄的树脂，以保证铺贴时胶液能完全浸渍修理补片。 Apply 80% ±5 % of mixed resin and keep a small amount of resin to apply a thin layer of resin to the repair area prior to lay-up and to wet out any dry areas of the fabric. （7）将裁剪好的纤维干布放在胶液上，确保纤维干布没有褶皱。 Place the cut dry fabric on the resin and make sure that the fabric is free from wrinkles. （8）使用干净的刮板、硬辊或刷子轻轻地浸渍纤维干布并排除气泡。 Use a clean spatula, hard roller or brush to gently impregnate the fabric and to remove trapped air. 注意：小心不要扭曲或损坏纤维干布。 Note: Take care not to distort or damage the fabric. 注意：如果织物上仍有未浸渍区域，则局部添加额外的树脂。 Note: If any dry areas of fabric remain add additional resin locally. （9）静置 5 min，排除材料中的部分空气。 Leave for 5 minutes to allow the entrapped air to escape. （10）在织物上覆盖第二张隔离膜，也称作绘图膜。 Put the second piece of parting film referred also as drawing film over the fabric. （11）使用刮板或硬辊轻轻地挤出空气。 Gently squeeze out the entrapped air using a squeegee or hard roller. 注意：此时不要将多余的树脂挤出。 Note: Do not work out excess resin at this time. 注意：确保将树脂涂到织物边缘并使其充分湿润。 Note: Make sure to apply the resin to the edge of the fabric and impregnate it thoroughly. （12）如果仍在织物中看到未浸渍区域，请添加更多的树脂，并重新从步骤 2 开始。 If any dry areas can be seen in the fabric, add more resin and restart from step 2. （13）当修理补片夹在隔离膜之间时，根据每个修补层所需的尺寸将浸渍的织物裁剪成相应尺寸，且标明经线（0°）方向，标注方式请参见图 4-1-1。 Whilst still sandwiched between the parting films, cut the impregnated fabric to the required dimensions for each repair ply. Identify the warp (0°) direction, refer to Fig.4-1-1.

任务名称 Task Topic	湿铺层制备 Wet-layup repair patch preparation
步骤 Steps	规章 / 指令 Regulation/Instruction
3. 湿铺层裁剪 Wet laying-up repair patch preparation	注意：对于非圆形补丁，请确保补片角有 0.5 in（12 mm）的半径。不允许有方形角落。小尺寸损伤 / 修补可归纳为小半径圆形损伤。 Note:　For none-circular repairs make sure that corners have a radius of 0.5 in. (12 mm). Square corners are not allowed. Small damage repairs may be treated as reduced radii circular damages.
4. 完成并清洁 Completion and cleaning	完成此工序后，要及时关闭设备，清理工具和耗材。 To timely close after completion of the process equipment, cleaning tools and materials. 使用干净、干燥的布或海绵轻按修理补片，除去多余的气泡和树脂。 Remove any excess resin from the edges of the repair patch using a clean, dry cloth or sponge.

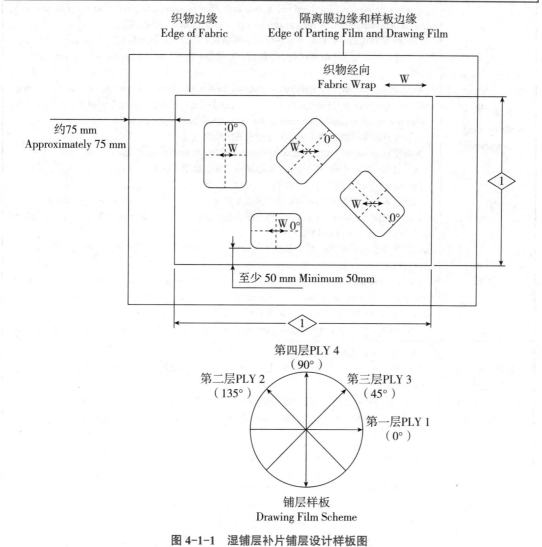

图 4-1-1　湿铺层补片铺层设计样板图

Fig.4-1-1　Wet laying-up repair patches orientations drawing scheme

任务名称 Task Topic	湿铺层铺贴 Wet-layup repair patch installation
步骤 Steps	规章 / 指令 Regulation/Instruction
1. 胶液准备 Adhesive preparation	胶液准备的施工操作可参考图 4-1-2。 The operation for preparing the adhesive solution can refer to Fig. 4-1-2. 注意：修理补片铺层角度应在具体章节查询。 Note: The orientation of the repair plies shall be as shown in the specific repair chapter.
2. 铺层准备 Laminating preparation	（1）使用刷子或刮板涂覆一层较薄的胶液于粘接或填充芯层修复区域。只需使用最少量的胶液浸润表面即可。 Use a brush or spatula to apply a thin coat of laminating resin to the laminate bond or filled core repair area. Use a minimum amount of resin to wet the surface. （2）如果修理区域存在已开口的蜂窝芯材，需向胶液中加入增稠剂（质量含量大约为5%）。 If repairing an area with open cell honeycomb core exposed, add thickening agent to laminating resin (approximately 5% by mass) to give a non-slumping consistency. （3）将覆盖在第一层修理补片一侧的分离膜取下。以与纤维方向成45°的角度剥离。使用刷子或抹刀均匀地涂抹浓稠树脂于与蜂窝芯材接触的修理表面。涂抹 0.02~0.1 g/ cm² 的浓稠树脂。 Remove the parting film from one surface of the first ply. Peel at 45° to the fiber direction. Apply uniformly the thickened resin with a brush or spatula over the exposed surface of the first ply which will be in contact with the open honeycomb core. Apply 0.02 to 0.1 g/cm² of resin. 注：这种浓稠树脂主要用于芯材与蒙皮表面胶接形成胶瘤。 Note: This thickened resin is used to form fillets for the core to skin joint. （4）将已揭掉一面隔离膜的第一层修理补片放在修理区域上。 Put the exposed surface of the first ply on the repair area. 注意：修复层的经向需与修理补片修理方案设定的方向对齐，参考图 4-1-3 和图 4-1-4。 Note: Align the warp direction of the repair ply with the direction required by the repair scheme according to Fig.4-1-3 and Fig.4-1-4. （5）使用刮板或硬辊，小心地挤压层合板，以消除皱纹和空气。从中心向边缘施工，避免将树脂挤出。 Use a squeegee or hard roller to remove wrinkles and entrapped air by carefully squeezing the laminate. Work from the center of the ply to the edges. Avoid removal of the resin. （6）将修理补片剩下的一层隔离膜剥离，剥离角度与纤维方向成45°。 Remove the parting film from the upper surface of the repair ply. Peel at 45° to the fiber direction.
3. 湿铺层补片安装 Wet laying-up repair patch preparation	（1）从已安装的补片上部表面45°撕离分离膜。 Remove the parting film from the upper surface of the installed repair ply peel at 45° to the fiber direction. （2）使用相同铺层基准依次铺贴修理补片。重复此步骤，直到所有修补层都铺贴完毕。 Replace one damaged ply by one repair ply with respect of basic orientation lay-up. Repeat steps until all the repair plies are laid up.
4. 预压实 Debulking	补充步骤：预压实，用下述材料覆盖修理区域： Supplementary step: Debulking, cover the repair area with the following vacuum equipment: （1）有孔隔离膜、玻璃纤维布、真空袋。 Perforated parting film, glass fabric cloth, vacuum bag. （2）施加真空压力到 0.8 bars (11.6 psi)，持续 5 min。 Apply a vacuum of 0.8 bars (11.6 psi) for a period of 5 minutes.

任务名称 Task Topic	湿铺层铺贴 Wet-layup repair patch installation		
步骤 Steps	规章 / 指令 Regulation/Instruction		
4. 预压实 Debulking	（3）移除真空设备，继续进行铺叠。 Remove the vacuum equipment and continue with the lay-up. （4）真空袋是保证修理质量的重要因素。 Vacuum bags are essential for the repair quality. （5）压力可以压实铺层，提高致密度和层间粘接性能。 Pressure compacts the laminate, providing good consolidation and interlaminar bonds. （6）真空可吸出挥发物质和空气，减少孔隙，使树脂流动性能提升。 Vacuum draws out volatiles and trapped air, resulting in a low void content. Both help to improve resin flow.		
5. 完成并清洁 Completion and cleaning	完成此工序后，要及时关闭设备，清理工具和航材。 To timely close equipment after completion of the process, cleaning tools and materials.		

图 4-1-2　胶液称量方法

Fig.4-1-2　**Method for weighing adhesive**

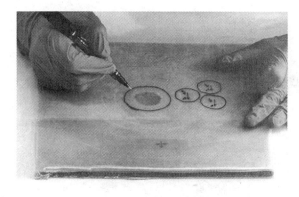

图 4-1-3　湿铺层绘图方法

Fig.4-1-3　**Method for wet lay-up layer drawing**

湿铺层制备

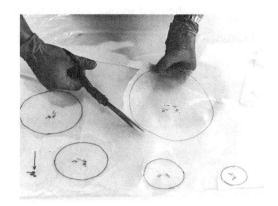

图 4-1-4　湿铺层补片裁剪方法

Fig.4-1-4　**Method for wet lay-up repair patches drawing**

湿铺层制备与铺贴型面练习题任务工卡
Wet Layup Preparation and Installation Exercise Task Card

（1）在制备复合材料湿铺层时，如何确定所需的树脂和纤维材料数量？有哪些注意事项？

How to determine the required amount of resin and fiber materials in preparing wet layup for composite materials? What are the precautions?

（2）如何处理在湿铺层制备和铺贴过程中可能出现的问题，如泡沫、气泡和褶皱等？

How to deal with potential issues that may arise during the preparation and layup of wet layup, such as foam, bubbles, and wrinkles?

（3）复合材料湿铺层的最终表面质量受哪些因素影响？如何评估其质量？

What factors affect the final surface quality of wet layup for composite materials? How to evaluate its quality?

湿铺层制备与铺贴操作工序卡

Wet Layup Preparation and Installation Operation Log

1. 工具、耗材、设备清点：

Inventory of tools, consumables, and equipment:

2. 型面清洁记录：

Surface cleaning record:

3. 湿铺层尺寸、铺层设计：

Wet layup layer size and lamination design record:

4. 配胶比例记录：

Adhesive mixing ratio record:

5. 湿铺层厚度、质量、平整度记录：

Wet layup layer thickness, quality, and flatness record:

6. 清点归还工具：

Tool inventory and return record:

操作总结：

Summary:

子任务 2　蜂窝修理芯塞安装
Subtask 2　Honeycomb Core Repair Plug Installation

【子任务解析 Subtask Analysis】

若损伤深度涉及蜂窝更换，则需执行蜂窝去除、蜂窝制作、蜂窝安装与除湿等相关程序。发现夹芯结构损伤后，首先要使用合适的检测方法对损伤部位进行检测，并确定损伤种类、损伤面积及损伤程度，在此基础上对损伤做准确、全面评估，然后查阅结构维修手册确定详细修理步骤。

If the depth of damage involves cell replacement, procedures such as cell removal, cell fabrication, cell installation, and dehumidification shall be performed. After the damage of sandwich structure is found, it is necessary to use appropriate detection methods to detect the damage site first, and determine the type, area and degree of damage. On this basis, it is necessary to make an accurate and comprehensive assessment of the damage, and then refer to the structural repair manual to determine the detailed repair steps.

通过本任务的训练，可以掌握全蜂窝深度的蜂窝修理流程，包括制作与安装的标准施工程序，也可迁移应用于不同损伤深度的蜂窝修理任务。

Through the training of this task, we can master the honeycomb core repair process of full damaged depth, including the standard procedure of fabrication and installation, and can also transfer the repair task applied to different damaged depths.

【子任务分解 Subtask Break-down】

（1）掌握蜂窝修理芯塞制作流程。

Master honeycomb core repair plug production process.

（2）掌握蜂窝修理芯塞安装流程。

Master honeycomb core repair plug installation process.

（3）掌握蜂窝修理芯塞固化与除湿方法。

Master honeycomb core repair plug curing and drying processes.

小任务 1　飞机上为何使用复合材料夹芯结构?
Subtask 1　Why are composite sandwich structures used on airplanes?

夹芯结构由两个薄且坚硬的外蒙皮面板和相对较厚、较轻的夹芯材料构成，蒙皮与芯材通过粘接层粘接在一起。芯材作为腹板，连接两侧外蒙皮，并为结构提供必要的刚度和强度，而外蒙皮作为翼缘法兰板，抵抗弯曲并为结构提供额外的强度。

The sandwich structure consists of two thin and rigid skin panels bonded to a relatively thick and light core material. This core material acts as a web, connecting the two skins and providing the necessary stiffness and strength to the structure. The skin panels, on the other hand, act as flanges, resisting bending and providing additional strength to the structure.

在航天领域中，夹芯结构可实现质量显著减轻，进而提高飞机燃油效率而降低运营成本。另外，夹芯结构还具有改善振动阻尼和降噪等优点，被广泛应用于船舶、汽车和建筑等其他领域。

The use of sandwich structures in aircraft manufacturing has led to significant weight savings, which in turn results in increased fuel efficiency and lower operating costs. In addition, sandwich structures offer other benefits such as improved vibration damping and noise reduction. Because of these advantages, sandwich structures are also used in other industries such as marine, automotive, and construction.

小任务 2　复合材料夹芯结构主要损伤形式
Subtask 2　Main damage forms of composite sandwich structures

复合材料夹芯结构主要损伤形式有冲击损坏和吸湿老化。目前，飞机制造商已在设计阶段考虑如何通过仔细选择材料和结构设计来防止昂贵的维护问题，甚至结构故障，其中使用适当的薄膜胶粘剂和外部密封系统，可有效防止夹芯结构的蒙皮由于水分进入或冻融循环而导致结构早期失效。

Main damage forms of composite sandwich structures are impact damage and hygroscopic aging. At present, aircraft manufacturers have considered in the design stage how to prevent expensive maintenance problems, even structural failures, by carefully selecting materials and structural design, even using the appropriate film adhesive and external sealing system, which can effectively prevent sandwich skin due to water may enter or freeze-thaw cycle and cause premature failure of the structure.

小任务3 蜂窝结构为何强调需执行除湿程序？

Subtask 3 Why is it emphasized to perform a drying procedure for honeycomb structures?

由于夹芯结构比层合板结构内部易产生积水，因此在进行修理之前，必须完全去除夹层结构内部的积水或湿气，否则，在固化时高温环境中产生的蒸汽压力会导致修理补片产生分层并降低其强度。清除夹芯结构内部的积水或湿气，通常优先选用电热毯加热去除，也可利用烘箱、烤灯和热风枪对夹芯结构进行烘干。

Since the sandwich structure is more prone to water generation than the laminate structure, the water or moisture must be completely removed from the sandwich structure before repair; Otherwise, the steam pressure generated in the high temperature environment during curing will cause the repair patch to delaminate and reduce strength. To remove the water or moisture inside the sandwich structure, electric blanket is usually preferred for heating removal, and oven, baking lamp and heat gun can also be used to dry the sandwich structure.

材料与工具清单——蜂窝安装
Materials and Tools List—Honeycomb Core Installation

工具 / 材料 Tools/Materials	图示 Illustration
蜂窝芯材 Honeycomb–core material 气动打磨机 Pneumatic grinding machine 防尘、防毒面具 Dustproof gas mask 清洗剂（丙酮、三氯乙烷等） Cleaning agent (acetone, trichloroethane, etc.) 橡胶手套 Rubber gloves	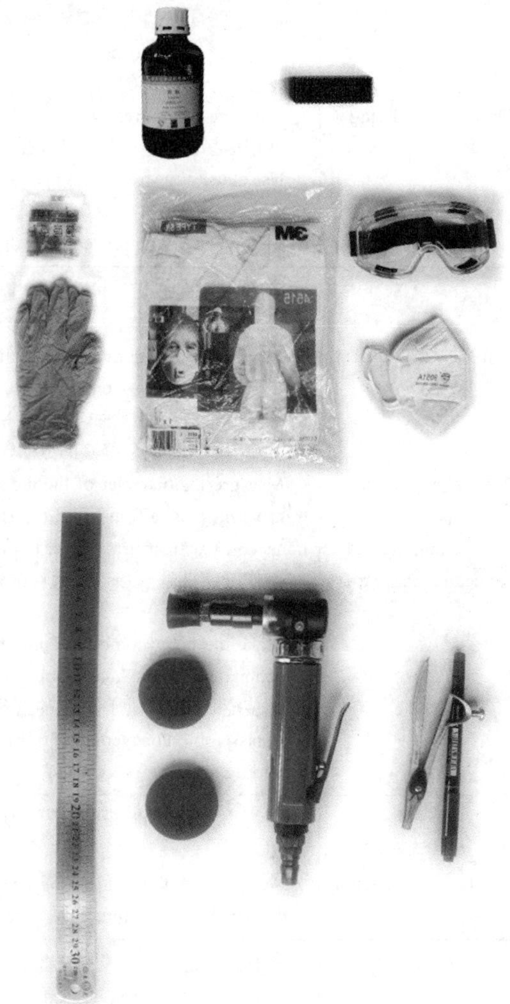

蜂窝安装任务工单
Honeycomb core Installation Task Card

任务名称 Task Topic	蜂窝安装 Honeycomb core Installation
步骤 Steps	规章 / 指令 Regulation/Instruction
1. 清洁与干燥 Cleaning and drying	用真空吸尘器清除所有灰尘； Remove all dust with vacuum absorber; 使用丙酮或经许可的溶剂润湿洁净的无尘擦拭纸擦拭； Use acetone or an approved solvent to moisten clean dust-free wipe paper; 用一块新的干燥的无尘擦拭纸再次擦拭修理区域切口； Wipe the repair area again with a new dry dust-free wiping paper; 使用洁净的无尘擦拭纸或使用不含油的压缩空气除去残留的溶剂； Use clean dust-free wiping paper or use oil-free compressed air to remove residual solvent; 按照除湿程序执行待修理件的干燥。 Follow the drying procedure to dry the parts to be repaired.
2. 高度修整与方向标记 Height trimming and direction marking	测量修理区域切口的深度，若蜂窝芯材被完全去除，则蜂窝芯塞的深度可在手册中查询； Measure the depth of the cut in the repair area. If the honeycomb core is completely removed, the depth of the honeycomb core plug can be queried in the manual. 测量修理区域切口的直径，可采用纸胶带或透明的塑料薄膜覆盖在切口处，描出切口形状，并用箭头标明原蜂窝的方向，如图 4-2-1 所示； Measure the diameter of the incision in the repair area, cover the incision with paper tape or transparent plastic film, trace the shape of the incision, and mark the direction of the original honeycomb with an arrow refer to Fig.4-2-1. 制作蜂窝芯塞模板，箭头对准蜂窝芯材的孔格平分线，并按照模板的形状切割出蜂窝芯塞。 Make the honeycomb core plug template. Point the arrow to align with the honeycomb core direction, and cut the honeycomb core plug according to the shape of the template. 预装配蜂窝，根据预装配的情况，用美工刀或打磨机修整蜂窝芯塞尺寸。 Preassemble honeycomb. Trim the size of honeycomb core plug with utility knife or sander according to the pre-assembly.
3. 填充层铺叠 Filling layers installation	根据损伤切口的直径剪下圆形胶膜（或湿铺层），视为填充层，将填充层铺叠于切口底部。 According to the diameter of the injured incision, the circular film (or wet layering layer) is cut and regarded as the filling layer, and the filling layer is laid on the bottom of the incision.
4. 安装修整 Installation and adjustment	根据蜂窝芯塞高度及周长剪下大小合适的胶膜，将胶膜贴合在蜂窝芯塞外围，剪去多余胶膜后将蜂窝芯塞小心放入孔洞，并确保蜂窝芯塞与原蜂窝接触面均有胶膜； Cut the appropriate size adhesive film according to the height and perimeter of honeycomb core repair plug. The film laminating surrounds the honeycomb core, then cut off excess film carefully and make sure that the honeycomb core plug is fully contacted with the original interface; 注意：在安装时，应当避免用裸手触摸胶膜，可用塑料薄膜包裹住胶膜，安装后，再将塑料薄膜抽出。 Note: During installation, do not touch the film with bare hands. Wrap the film with plastic film and pull out the plastic film after installation.

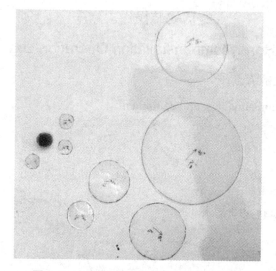

湿铺层

图 4-2-1　蜂窝安装芯格方向标记示意图

Fig.4-2-1　Schematic diagram of honeycomb installation with directional markings

蜂窝安装练习题任务工卡
Honeycomb Core Installation Exercise Task Card

（1）如何在安装过程中确保蜂窝芯材的平整度和对准度，不正确的对准会导致什么后果？

How can the flatness and alignment of the honeycomb core be ensured during installation, and what are the consequences of improper alignment?

（2）对蜂窝芯材的边缘填充材料的目的是什么，通常用于填充的材料是什么？

What is the purpose of potting the edges material of the honeycomb core, and what materials are commonly used for potting?

蜂窝安装操作工序卡
Honeycomb Installation Operation Log

1. 工具、耗材、设备清点：

Inventory of tools, consumables, and equipment:

2. 清洁与干燥记录：

Cleaning and drying record:

3. 蜂窝修理芯塞尺寸记录：

Honeycomb repair core plug size record:

4. 蜂窝修理芯塞芯格对齐记录：

Honeycomb repair core plug core-grid alignment record:

5. 填充层尺寸与铺层记录：

Fill layer size and layup record:

6. 清点归还工具：

Tool inventory and return record:

操作总结：

Summary:

任务4 损伤修补考核表

姓名		班级		
评价维度	分值／分	自评（30%）	互评（30%）	师评（40%）
素养（20%）				
1. 材料清点齐全	5			
2. 任务书自学情况	5			
3. 安全文明操作及6S	10			
技能（60%）				
1. 配胶操作规范性 （配胶比例、配胶气泡程度）	10			
2. 湿铺层设计正确性 （数量、尺寸设计）	20			
3. 湿铺层安装操作规范性 （气泡、平整性、含胶量计算）	20			
4. 蜂窝安装操作规范性 （尺寸、芯格对齐）	10			
总结报告（20%）				
1. 训练总结的完整性	10			
2. 个人反思与拟订后续训练计划	10			
总计	100			
任务完成情况	提前完成			
	准时完成			
	滞后完成			

封装、固化与修理后处理
Encapsulation, Curing and Post-repair Treatment

【任务情境 Task Scenario】

你作为 WS 航空公司的航线大修人员，持有复合材料部件维修授权书，现接到一件待修复合材料部件，需进行真空封装与固化。在这个任务中，你的主要职责包括制备真空袋、进行真空封装并根据树脂系统的特性和制造工艺要求进行固化，并检验修理质量。

As a line maintenance technician of WS Airlines and authorized to repair composite material components, you have received a composite material component for repair which requires vacuum sealing and curing. Your main responsibilities in this task include preparing the vacuum bag, performing vacuum sealing, and curing according to the characteristics of the resin system and manufacturing process requirements, and inspect the repair quality.

【任务解析 Task Analysis】

封装、固化并检验是进行复合材料结构损伤维修工作的最后一步。完成本项目训练后，应实现以下目标：

Encapsulation, curing and inspection is the last step of damage repair for composite structures. After completing the training of this project, the following goals should be achieved:

知识目标 Knowledge Objectives

（1）熟悉复合材料真空封装固化的基本概念、工艺流程和常用工具材料。

You should be familiar with the basic concepts, process flow, and common tools and materials used in vacuum sealing and curing of composite materials.

（2）熟悉不同树脂系统的特性和固化条件，以及不同复合材料部件的制造工艺要求。

You should also understand the characteristics and curing conditions of different resin systems, as well as the manufacturing process requirements for different composite material components.

（3）理解真空封装、固化的原理和作用机制，以及各种固化问题的处理方法。

You should have a good understanding of the principles and mechanisms of vacuum sealing and curing, as well as the methods for handling various curing issues.

（4）理解修理后处理中针对漆层恢复等表面恢复技术的处理方法。

You should have a good understanding of the post-repair treatment methods for surface restoration, including techniques for restoring paint layers and other surface aspects.

能力目标 Ability Objectives

（1）熟练复合材料封装前的准备工作，按照维修手册正确裁剪辅助材料尺寸，按特定顺序正确放置辅助材料铺层。

You should be proficient in the preparatory work before composite material packaging, such as cutting auxiliary materials to the correct size according to the maintenance manual and placing them in the correct order.

（2）掌握封装操作流程，保证真空袋气密性良好。

You should master the packaging process and ensure good vacuum airtightness.

（3）掌握根据维修情况确定固化的方法，正确完成维修区域的固化流程。

Master the method of determining the curing process according to the repair situation, and correctly complete the curing process of the repair area.

素质目标 Emotion Objectives

（1）具有严谨的工作态度和责任心，认真执行各项操作规程，确保工作质量和安全。

You should have a rigorous work attitude and sense of responsibility, carefully executing all operation procedures to ensure work quality and safety.

（2）具有团队协作精神和沟通能力，与其他修理人员、工程师和客户进行有效沟通，确保工作顺利进行。

You should have a team spirit and communication skills, effectively communicating with other technicians, engineers, and customers to ensure smooth work progress.

（3）具有学习和探索精神，持续提高专业知识和技能水平。

You should have a spirit of learning and exploration, continuously improving your professional knowledge and skill levels.

辅助材料的识别与使用
Recognition and usage of auxiliary materials

辅助材料的铺放程序
Laying-up sequence of auxiliary materials

封装固化
Encapsulation, Curing

固化制度
Curing methods

质量控制
Quality control

修理后处理
Post-repair treatment

子任务 1　封装和预压实
Subtask 1　Encapsulation and Debulking

【子任务解析 Subtask Analysis】

封装是指在未固化的零件上使用真空袋薄膜覆盖并密封其边缘，从而可对其进行抽真空，利用真空压力或外界压力使需要固化的材料结合在一起。真空袋需要良好的密封以防止由于真空泄漏而引起的失压。在完成对蜂窝损伤的修补后，对维修区进行封装，旨在为后续固化做准备。真空封装前需将辅助材料按照指定的封装顺序铺置。

Encapsulation is the use of an impermeable membrane on an uncured part to cover and seal its edges so that it can be evacuated and the parts to be glued together using atmospheric pressure. The vacuum bag needs to be well sealed to accommodate loss of pressure due to leakage. After repairing the honeycomb damage, encapsulate the repair area in preparation for subsequent curing. Before vacuum packaging, the auxiliary materials should be laid in the specified packaging order.

通过该任务的训练，需掌握正确的真空封装方式，按照标准顺序完成辅助材料的铺放，熟练操作封装具体流程，并保证真空袋气密性良好。

Through the training of this task, we can master the correct vacuum packaging method, and complete the laying of auxiliary materials according to the standard order. Proficient in the specific process of packaging, and ensure good vacuum tightness.

【子任务分解 Subtask Break-down】

（1）掌握真空袋制作方法，并保证气密性。

Master the vacuum packaging method, and ensure air tightness.

（2）理解并掌握按照标准顺序铺放辅助材料的方法。

Understand and master the method of placing auxiliary materials in standard order.

（3）熟练操作封装具体流程。

Familiar with the specific process of packaging.

小任务 1　常见真空封装形式有哪些？如何选择合适的封装形式？

Subtask 1　What are the common forms of vacuum packaging? How to determine the correct packaging form?

（1）表面封装 Surface packaging

复合材料表面真空封装，也称为单面封装，是指直接在复合材料构件表面施加真空袋薄膜与密封胶带，在真空泵的作用下将袋内空气排除干净，形成密封真空袋。此方法作为首选的封装方法，要求维修部件尺寸足够大，且可以在维修的一侧放置真空袋。

Surface vacuum bagging for composite materials, also known as single-sided vacuum bagging, involves the application of a vacuum bag film and adhesive tape directly onto the surface of the composite component. The vacuum pump is utilized to evacuate all air from the bag, creating a hermetically sealed vacuum environment. This method is the preferred encapsulation technique, which necessitates the repaired component to be of adequate size and allows the vacuum bag to be placed on one side during the repair process.

（2）整体封装 Overall packaging

整体真空封装是一种技术手段，通过在整个复合材料构件上创造真空环境，以确保其结构完整性和性能稳定性。该方法通常涉及在复合材料构件表面覆盖多层封装材料，随后将整个构件放入真空袋中，利用真空泵抽取气体以形成负压，从而使复合材料构件在高温下固化。

Overall vacuum sealing of composite materials is a technique that creates a vacuum environment throughout the entire composite material component to ensure its structural integrity and performance stability. Whole-vacuum sealing typically involves covering the surface of the composite material component with a hierarchical sealing material, then placing the entire component in a vacuum bag, extracting gas through a vacuum pump to create negative pressure, and allowing the composite material component to automatically cure at high temperatures.

使用这种封装方式，全部真空压强作用于整个工装面积。工装强度必须符合要求，以避免在固化温度下发生蠕变变形。整体封装通常不适用于大型控制面，如由复合材料夹层壁板、梁和肋组成的方向舵结构，每个壁板组装前是由热压罐单独胶接的，通常不能承受封装过程中的压力，若采用整体封装方式，则压力可能将壁板压塌，因此针对带胶接结构的大型控制面，所有维修都必须采用局部表面封装的办法，或将壁板从组件中拆卸下来，再对壁板单独进行整体封装。

With this packaging, the full vacuum pressure is applied to the entire tooling area. Tooling strength must be higher to avoid creep deformation at curing temperature. A problem arises when this method is used for large control surfaces, such as rudders composed of composite cored panels and beams and ribs. If such a large assembly were encapsulated in one piece, the pressure could crush the panels. Each panel is individually glued by an autoclave prior to assembly, and the assembly is designed so that it cannot withstand the pressures designed for the encapsulation process. In such cases, all repairs must be done by partial surface encapsulation, or by disengaging the panels from the assembly and then encapsulating the panels individually.

小任务2　封装过程中需要哪些辅助材料？各辅助材料的功能是什么？
Subtask 2　What auxiliary materials are required in the packaging process, and what are their functions?

（1）剥离布主要用于分离工件与封装材料，可以吸收一部分胶液且便于固化后将封装材料从构件上剥离。

The peel ply is used as a separator. The resin can pass, but the bagging materials can be removed after curing from the repair plies.

（2）有孔隔离膜用于控制树脂流出量。

The perforated parting film is used to control the resin flow.

（3）吸胶布吸收多余的树脂，防止溢胶。

The bleeder cloth absorb the excess resin to prevent overflow.

（4）无孔隔离膜多为聚四氟乙烯薄膜，是一种起隔离作用的薄膜材料，防止胶接结构固化后与模具或盖板及其他辅助材料粘贴。

Non-perforated parting films are commonly made from polytetrafluoroethylene films. They serve as isolating films, preventing the bonded structure from adhering to the mold, cover plate, or other auxiliary materials after curing.

（5）透气毡可以将固化过程中产生的多余气体导入真空管道中，保证复合材料的成型质量。由于透气毡是气体向外流动的通道，其尺寸必须大于无孔隔离膜。

Breather serves as a conduit for directing excess gas produced during the curing process into the vacuum pipeline, thereby ensuring the quality of composite material molding. Due to its role as a gas outflow pathway, the dimensions of the breathable felt must be greater than that of the non-perforated parting film.

（6）均压板为具有一定的柔性的金属薄板，可用于提高制件表面平整度，改善尺寸精度并减少固化过程的铺层滑移。

A pressure plate is a flexible metal thin plate that can be used to improve the surface flatness of the workpiece, enhance dimensional accuracy, and reduce layer slippage during the curing process.

（7）真空薄膜应具有较好的强度、延展性、耐温性、耐磨性和韧性。使用时，用密封胶带将成型中的构件密封在模具上。

The vacuum film should possess good strength, ductility, temperature resistance, wear resistance, and toughness. When in use, the molded components are sealed to the mold with putty tape.

（8）密封胶带应该具有常温下良好的黏性、高温下密封性好、固化后易清理和储存时间长等特点。

The sealant tape should have good adhesiveness at room temperature, good sealing performance at high temperatures, easy cleaning after solidification, and long storage time.

（9）压敏胶带主要起定位和固定的作用。

Pressure-sensitive adhesive tape is mainly used for positioning and fixing.

小任务 3　封装工序通用程序原则是什么，以及辅助材料放置的注意事项有哪些?

Subtask 3　What are the general principles of the encapsulation process, and what are the precautions for placing auxiliary materials?

复合材料真空封装技术是一种常用于制造和修理高性能复合材料部件的加工方法。主要将预浸有树脂的纤维织物铺放于一个密封的真空袋内，然后将真空袋在加热条件下固化，使树脂发生交联固化反应达到零部件所需的形状和结构。封装工序包括材料铺叠、制备真空袋、气密性检验、加热固化等步骤。

Vacuum sealing technology for composite materials is a commonly used processing method for manufacturing high-performance composite material components. It mainly involves stacking pre-impregnated composite material layers with resin in a special bag, then evacuating the air in the bag to achieve a vacuum state, and then sealing the bag and curing the resin under heating conditions to form the desired shape and structure. The sealing process includes stacking composite materials, preparing a vacuum bag, air tightness inspection, heating and curing, and other steps.

辅助材料的放置顺序需要按照手册要求执行，以保证修理件符合质量要求，放置顺序主要包含以下注意事项。

The placement sequence of auxiliary materials needs to be carried out according to the manual requirements to ensure that the repaired parts meet the quality requirements, and the placement order mainly includes the following considerations.

（1）通常需要先放置热电偶，在修理区域外周均放至少 3 个热电偶。

Generally, thermocouples should be positioned before the encapsulation process, with a minimum of 3 thermocouples evenly placed around the periphery of the repair area.

（2）通常依次铺放有孔隔离膜、吸胶布、无孔隔离膜、电热毯和透气毡，透气毡放置在电热毯的上方，防止对真空袋造成损伤。

Generally, the sequence of placement includes perforated isolation film, absorption cloth, non-perforated parting film, electric heating blanket, and breather. The breather is placed above the heating blanket to prevent damage to the vacuum bag.

（3）在维修区域外侧的位置放置真空表和真空接头，再在修理区域外围粘贴密封胶带，随后施加真空袋薄膜将其密封。

Place the vacuum gauge and vacuum connector on the outside of the repair area, and then apply sealant tapes around the periphery of the repair area. Subsequently, apply the vacuum bag film to seal it.

（4）对于平板件修理，可以在透气毡之上、电热毯之下，放置一块均压板。

For panel repairs, a caul plate can be placed above the breather and below the electric heating blanket.

针对蜂窝夹芯结构，根据蜂窝夹芯结构的厚度不同，放置电热毯和热电偶的方式存在差异。

Regarding honeycomb sandwich structures, the placement of heating blankets and thermocouples varies depending on the thickness of the honeycomb layers.

（1）如果芯塞的厚度 ≤ 0.5 in，则只在构件外侧铺放一层电热毯即可，并至少在维修区域设置两个单独的热电偶。

If the thickness of the core plug is ≤ 0.5 in., only lay the heating blanket on the outside, and put at least two separate thermocouples in the maintenance area.

（2）如果芯塞厚度 > 0.5 in，并且两面都具备施工条件时，则需在构件两侧均铺放电热毯以及放置两个单独的热电偶，除此之外，为了监控芯材内部的固化质量，还需要在芯材中心放置一个热电偶。

If the thickness of the core plug is >0.5 in., and both sides of the component are reachable, electric heating blankets should be placed on both sides, along with two separate thermocouples. Moreover, to monitor the curing quality within the core material, an additional thermocouple should be positioned at the center of the core material.

（3）如果芯塞厚度＞0.5 in，但构件只有单面具备施工条件时，则只需要在此侧铺放电热毯，并在蜂窝芯材的芯格中至少放置两个与芯格下部的维修材料相接触的热电偶，并且使用压敏胶带固定。

If the thickness of the core is >0.5 in., but only the outer side can be reached, then lay a heating blanket on the outside, and put at least two thermocouples into the core plug cavity, so that the thermocouples are in contact with the maintenance material at the lower part of the hole, and the probe head of the thermocouple needs to be fixed with tape.

小任务 4　双真空压实技术是指什么？
Subtask 4　What is meant by Double Vacuum Debulking technology?

双真空压实（Double Vacuum Debulking，DVD）技术是一种先进的复合材料制造技术，利用双真空室制备预固化复合材料修理补片，从而实现将复合材料预浸料或干预浸料进行同时压实和固化。第一个真空室通常用于去除空气和挥发性成分，第二个真空室用于在压力下固化材料。图 5-1-1 给出了 DVD 工艺的详细封装方法。

Double Vacuum Debulking (DVD) technology is an advanced composite manufacturing technique that uses two vacuum chambers to simultaneously consolidate and cure pre-impregnated or dry preform composite materials. The first vacuum chamber is typically used to remove air and volatile components, while the second vacuum chamber is used to cure the material under high pressure. This technique significantly improves the efficiency and quality of composite material manufacturing and is suitable for producing parts of various shapes. Fig.5-1-1 shows the detailed packaging method of the DVD process.

DVD 修理工艺主要用于复合材料铺层大于 6 层的损伤修理，因为过厚的复合材料修理铺层会导致固化过程中层合板含有过多的空隙，而国外航空公司一般要求修理件孔隙率小于 4%。DVD 工艺与单真空袋工艺主要的区别是在单真空袋封装的基础上，增加了 DVD 盒子，并在盒子外增加了一层密封真空袋，在不降低修理区域绝对真空度的前提下，提升了树脂内气泡与树脂的相对运动速度，有利于气泡的排出。相比单真空袋修理方法，DVD 工艺的主要优点是在不降低绝对压力的情况下，降低了修理区域的相对压力，从而提高了该区域树脂流动性，有利于气泡的排除而提高修理质量。

The DVD repair process is mainly used for damage repair with a composite layer of more than 6 layers, because too thick composite repair layer will cause the curing process to contain too much space, and foreign airlines generally require that the porosity of the repair parts is less than 4%. The main difference between the DVD process and the single vacuum bag process is that on the basis of the single vacuum bag package, the DVD box is added, and a layer of sealed vacuum bag is added outside the box, which improves the relative motion speed of the bubble and the resin in the resin without reducing the absolute vacuum degree of the repair area, which is conducive to the discharge of the bubble. Compared with the single vacuum bag repair method, the main advantage of the DVD process is that the relative pressure in the repair area is reduced without reducing the pressure, thus improving the resin flow in the area, which is conducive to the elimination of bubbles and improve the repair quality.

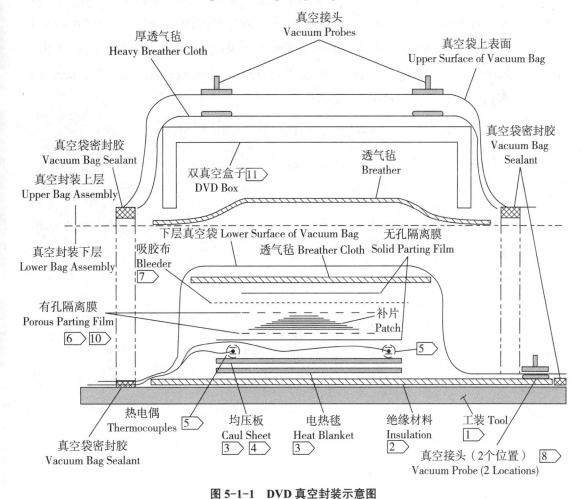

图 5-1-1　DVD 真空封装示意图

Fig.5-1-1　Schematic Diagram of DVD Vacuum Encapsulation

目前，民用航空复合材料薄板结构修理常采用单真空袋热补仪修理，该方法操作相对简单，对于修理铺层小于 6 层的层合复合材料修理效果较好。而当修理铺层数大于 6 层时，修补片的孔隙率一般会超过 5%，超出了国外飞机制造商最大 4% 孔隙率的要求。复合材料修补片如果存在空隙，对修补片与母片的贴合质量产生影响，容易引起修补片与母片贴合面上的应力集中，这对修复强度影响很大，而 DVD 修理工艺可实现大厚度层合板及其补片修理后的低孔隙率的目标。

Currently, the repair of thin composite structures in civil aviation mostly uses the single vacuum bag hot bonding repair method, which is relatively simple to operate and has better repair effects for repairing laminated composite materials with fewer than 6 layers. However, when the number of layers repaired exceeds 6, the porosity of the repair patch generally exceeds 5%, which exceeds the maximum 4% porosity rate requirement of foreign aircraft manufacturers. If there are voids in the composite repair patch, it can affect the fit between the patch and the parent plate and easily cause stress concentration on the fit surface between the patch and the parent plate, which greatly affects the repair strength. In contrast, the DVD repair process can significantly reduce the porosity of thick laminates and repair patches.

材料与工具清单
Materials and Tools List

材料 / 工具 Materials/Tools	图示 Illustration
热电偶 Thermocouples 压敏胶带 Pressure sensitive tape 电热毯 Heat blanket 透气毡 Breather cloth 两个真空接头 Two vacuum probes 真空袋 Vacuum bag 密封胶带 Sealant tape 剪刀 Scissor 有孔隔离膜 Perforated parting film 无孔隔离膜 Non-perforated parting film 均压板 Caul plate 钢板尺 Steel ruler	

封装任务工单
Encapsulation Task Card

任务名称 Task Topic	封装 Encapsulation
步骤 Steps	规章 / 指令 Regulation/Instruction
1. 放置有孔隔离膜 Perforated parting film laying up	放置一层有孔隔离膜，大于修理区域 1.0 in。 Put a layer of perforated parting film that is a minimum of 1.0 in. larger all around the repair area, above the repair area.
2. 放置热电偶 Thermocouple installation	如果在高于室温的温度下修理，将热电偶放在蜂窝修理程序中指定的位置，并将热电偶连接到适用的记录仪上。 If repairing at a temperature above room temperature, place the thermocouple in the position shown in the instructions for the vacuum bag for the honeycomb repair with the oven core and connect the thermocouple to a suitable logger.
3. 放置吸胶层 Bleeder installation	放置一层玻纤布来充当吸胶层，大于有孔隔离膜 2.0 in。 Put a layer of dry peel ply or glass fabric cloth above 2 in. more than the perforated parting film.
4. 放置无孔隔离膜 Solid parting film laying up	将一层无孔隔离膜放在吸胶层上，四周尺寸小于吸胶层 0.5 in。 Put a layer of solid parting film over the surface bleeder that is 0.5 in. less all around than the surface bleeder.
5. 放置透气毡 Breather cloth installation	在修补区上方放一层与物体表面大小相同的透气毡。真空封装内部辅助材料放置规范可参考图 5-1-2。 Place a layer of breather the same size as the surface of the object over the repair area The placement specifications for auxiliary materials inside vacuum packaging can be referred to Fig.5-1-2.
6. 真空袋密封 Vacuum bag encapsulation	用密封胶带密封真空袋。 Seal the repair with vacuum bag material. 将真空管和真空表的基座放置在一定位置。 Put the vacuum line and gauge base in place. 如果需要，在铺层处使用一圈密封胶条。 If necessary, apply vacuum sealant around the layup. 在真空管和真空表基座位置处的真空袋上切一个小开口。 Place the vacuum line in the vacuum bag and cut the vacuum bag attached to the base of vacuum gauge. 安装真空管和真空表到真空袋内的基座上。 Install the vacuum line and vacuum gauge, and record the indicator.
7. 气密性检验 Vacuum Integrity Testing	检查真空袋是否漏气。 Inspect for potential air leakage in the vacuum bag.
	注释：真空袋如果有漏气现象，会导致修理孔隙率增加以及随后的胶接失效。 Note: Air leakage in the vacuum bag can result in an elevated repair porosity and subsequent adhesive failure. 在修理区域施加最少为 22 英寸汞柱（558.8 毫米汞柱）的压力。 Apply a minimum pressure of 22 in. (558.8 millimeters) of mercury to the designated repair area.
	移除真空源。 Discontinue the vacuum source.
	监控真空表，5 min 后，真空表上改变的值必须小于 5.0 英寸汞柱（127 毫米汞柱）。 Monitor the vacuum gauge for a period of 5 minutes, ensuring that any observed change in the gauge reading is less than 5.0 in. (127 millimeters) of mercury.

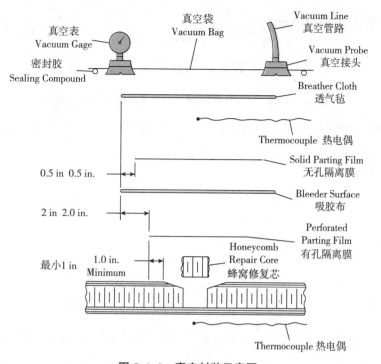

图 5-1-2　真空封装示意图

Fig.5-1-2　Vacuum bag encapsulation scheme

封装

真空封装练习题任务工卡

Vacuum Encapsulation Exercise Task Card

（1）真空袋封装的主要目的是什么？

What is the main purpose of vacuum bag?

（2）在真空袋封装的过程中，需要注意哪些常见的问题，以及如何避免它们？请结合实践练习总结归纳。

What are the common problems encountered during vacuum bag encapsulation and how to avoid them? Please sum up with practice.

（3）在真空袋封装中，如何处理异形复合材料？

How are composite materials with different shapes handled during vacuum bag encapsulation?

封装操作工序卡

Vacuum Encapsulation Operation Log

1. 工具、耗材、设备清点：

Inventory of tools, consumables, and equipment:

2. 辅助材料型号与尺寸记录：

Auxiliary material model and size record:

3. 辅助材料放置顺序记录：

Auxiliary material placement sequence record:

4. 真空度记录：

Vacuum level record:

5. 清点归还工具：

Tool inventory and return record:

操作总结：

Summary:

子任务2 固化
Subtask 2 Curing

【子任务解析 Subtask Analysis】

固化是指将封装好的预浸料或湿铺层材料通过合理的加温处理，使树脂的化学反应在经过控制的温度条件下进行，从而达到最佳的粘接效果并缩短固化时间。常见的固化加温设备有加热灯、烘箱、电热毯、热风枪等。对于湿铺层的固化，可采用室温固化或加温固化。加温固化过程中，不同的树脂对应要求的固化时间和温度均不相同，应参照产品的使用说明进行设置。

Curing refers to the controlled heating process of pre-impregnated materials or wet lay-up materials after they have been encapsulated, enabling the chemical reaction of the resin to occur under controlled temperature conditions, thereby achieving the best bonding effect and shortening the curing time. Common curing equipment includes heating lamps, ovens, electric blankets, heat guns, and so on. For wet lay-up curing, it can be carried out at room temperature or with heating. During the heating curing process, different resins require different curing times and temperatures, and these should be set according to the product's instructions for use.

通过该任务的训练，学生可掌握固化参数的设置以及熟练操作固化设备的具体流程。

Through the training of this task, students can master the setting of curing parameters and the specific process of operating curing equipment.

【子任务分解 Subtask Break-down】

（1）了解固化的定义、分类及常见的固化设备。

Understand the definition, classification, and common curing equipment.

（2）掌握湿铺层的加热固化方法。

Master the heating and curing method of wet layup.

（3）熟练操作修补芯塞的固化流程。

Proficient in the curing process of the repair core plug.

小任务 1　固化的定义是什么？有什么分类与方式？如何确定固化方式？

Subtask 1　What is the definition of curing, and what are the classifications and methods? How to determine the curing method?

固化是指通过光、热、辐射或化学添加剂等的作用，使热固性树脂经不可逆的化学反应完成交联的过程。

Curing refers to the process in which thermosetting resins are cross-linked through irreversible chemical reactions through the action of light, heat, radiation or chemical additives.

固化根据不同温度可分为室温固化和加温固化。

Curing can be divided into room temperature curing and heating curing according to different temperatures.

室温固化用于不重要的蜂窝夹芯结构的维修。室温固化维修中，为了缩短树脂固化的时间，往往也采用加热的方式来进行。通常，室温固化加温不超过 150 ℉，为了提高构件的力学性能，可使用真空袋进行加压。

Room temperature curing is used for repairs of unimportant honeycomb sandwich structures. In room temperature curing maintenance, in order to shorten the curing time of the resin, heating is often used to achieve its purpose. Typically, room temperature curing with heating does not exceed 150 ℉. In order to improve the mechanical properties of the component, a vacuum bag can be used for pressure application.

SRM 推荐的加温固化有三种温度，分别是低温固化（200~230 ℉）、中温固化（250 ℉）和高温固化（350 ℉）。

There are three temperatures recommended for heat curing in the SRM, which are low temperature curing (200–230 ℉), medium temperature curing (250 ℉), and high temperature curing (350 ℉).

固化根据不同的固化对象可分为层压结构固化和夹芯结构固化。

Curing can be divided into laminate structure curing and sandwich structure curing according to different curing objects.

固化设备有热压罐、烘箱、热补仪和电热毯等，如图 5-2-1 所示。

The curing equipment includes autoclave, curing oven, heat patch instrument and heating blanket, etc. as shown in Fig.5-2-1.

图 5-2-1　固化设备（从左到右依次为热压罐、烘箱、热补仪）

Fig.5-2-1　Curing equipment (from left to right: autoclave, curing oven, and heat compensation device)

热压罐固化要求高质量的工装，以提供每个零件的特殊形状和承受树脂固化和胶接过程中的温度，因此除非损伤非常严重，否则仅用电热毯进行固化修理即可。

Autoclave curing requires high quality tooling to provide the specific shape of each part and to withstand the temperatures of the resin curing and gluing process, so unless the damage is very severe, curing repairs with a heating blanket alone will suffice.

注意：当使用加热设备时，应遵守制造商的操作说明。在固化周期内保持最低 22 英寸汞柱的真空度，一般维修以热补仪和电热毯固化为主。

Note: When you use heating equipment, obey the manufacturer's operation instructions. Keep a minimum vacuum of 22 in. of mercury during the cure cycle. The usual procedure for these repairs is to cure the repair with a heat blanket.

小任务 2　影响固化结果的因素有哪些？
Subtask 2　What are the factors that affect the curing result?

（1）固化温度和时间 Curing temperature and time

固化温度和时间根据材料而确定，是固化过程中的重要因素。固化温度和时间决定了零件的固化度。

The curing temperature and time are determined according to the material and are an important factor in the curing process. The curing temperature and time determine the degree of cure of the part.

固化温度可通过热电偶监控。使用三个或更多热电偶监控夹具和零件的温度，将热电偶放在升温最快和最慢的地方，所有热电偶必须接触零件。

Cure temperature can be monitored by thermocouple. Use three or more thermocouples to monitor the cure cycle. Put the thermocouples at the locations where the temperature will increase the fastest and slowest. All of the thermocouples must touch the part.

注意：固化时间不包括铺层材料达到固化温度范围所需时间。固化时间通常指在进行热固化、化学固化过程中，零件在特定温度范围内的保温时间，这个时间会根据具体的工艺和材料要求有所不同，以确保树脂、胶粘剂或其他材料充分固化或反应。在固化时间开始计时之前，所有热电偶都必须处于固化温度范围内。如果热电偶示数降低到固化温度范围以下，应将固化时间延长，直到热电偶达到所需固化温度范围。

Note: Cure time does not include the time necessary for the layup and the part to get to the cure temperature range. Curing time typically refers to the duration during which a component is held within a specific temperature range during a thermal or chemical curing process. This time varies based on specific processes and material requirements to ensure thorough curing or reaction of resins, adhesives, or other materials. All of the thermocouples must be in the cure temperature range before the cure time starts. If a thermocouple indication decreases below the cure temperature range, extend the cure cycle time by the time necessary to get the thermocouple into the cure temperature range.

（2）保温时间 Holding time

保温时间通过每根热电偶在保温平台上的时间计算。

The holding time is calculated by the time of each thermocouple on the holding platform.

（3）固化压力 Curing pressure

固化压力决定了零件内部致密性，是零件内部质量控制的关键因素。固化压力根据零件构型和固化方式确定。常见的固化压力有 0.6 MPa、0.3 MPa 和真空压力。

The curing pressure determines the internal compactness of the part and is a key factor in the internal quality control of the part. The curing pressure is determined according to the mechanism form and curing method of the part. Common curing pressures are 0.6 MPa, 0.3 MPa and vacuum pressure.

（4）升降温速率 Rate of heating and cooling

通过升温、降温速率控制变温过程中零件的固化反应和热均匀性。加热和冷却速率等于在任何 10 min 内由单个热电偶测量的温差除以测量所经过的时间。

The curing reaction and thermal uniformity of the part during the temperature changing process are controlled by the rate of heating and cooling; the rate of heating and cooling is equal to the temperature difference measured by a single thermocouple within any 10min divided by the time elapsed for the measurement.

小任务 3 完整的固化程序可分为哪几个过程？每个过程需要注意什么？

Subtask 3 What are the complete curing program can be divided into, and what should be paid attention to in each process?

完整的固化程序可以分为升温、保温和降温三个过程。在固化过程中要注意以下几点：

The complete curing procedure can be divided into three processes: heating, heat preservation and cooling. During the curing process, pay attention to the following points:

（1）需要遵守修理方案中关于蜂窝厚度限制的要求，蜂窝厚度限制保证了足够的热量能够穿透蜂窝芯塞以使树脂发生交联固化反应。特别是在结构件只有单侧蒙皮可以接近的情况下，蜂窝厚度的限制对修理质量尤为重要。

It is necessary to adhere to the honeycomb thickness limits specified in the repair plan. The honeycomb thickness limit ensures sufficient heat penetration into the honeycomb core plug for the resin to undergo crosslinking and curing reactions. This limitation is especially critical for repair quality, particularly in cases where the structural component can only be accessed from one side of the skin.

（2）如果采用真空袋，应保持 0.8 bar 的最低压力。

If a vacuum bag is used, a minimum pressure of 0.8 bar should be maintained.

（3）将修理区域的温度升高到树脂固化所需的温度，升温速率控制在 3 ℃ /min。

Raise the temperature of the repaired area to the temperature required for resin curing, and control the heating rate at 3 ℃ /min.

（4）使结构件在规定温度下依据树脂固化参数规定的时间进行交联反应。注意固化时间从热电偶指示已达到要求的固化温度时开始计算。

Facilitate the crosslinking reactions of the structural component at the prescribed temperature, in accordance with the resin curing parameters' specified time. Take note that the curing time should commence its calculation once the thermocouple indicates that the required curing temperature has been achieved.

（5）当保温结束时，要在保持设定压力的条件下，以 3 ℃ /min 的速率冷却。当温度降到 50 ℃（122 ℉）或更低时，解除压力。

When the heat preservation is over, it should be cooled at a rate of 3 ℃ /min while maintaining the set pressure. When the temperature falls to 50 ℃ (122 ℉) or lower, depressurize.

固化过程中，针对不同的时间段某些因素可能对维修质量产生的影响程度的不同，又人为地将固化过程按时间分为四个区，如图 5-2-2 所示。

During the curing process, the curing process is divided into four zones according to time according to the different degree of influence on the maintenance quality in different time periods, as shown in Fig.5-2-2.

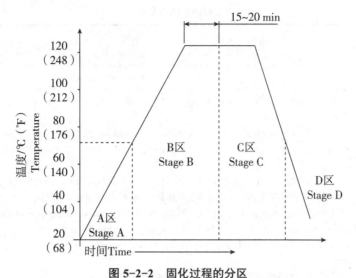

图 5-2-2　固化过程的分区

Fig.5-2-2　Zoning of the curing process

注释 Notes：

A 区：若发生固化温度骤降等问题，首先检查真空袋有无泄漏以及设备有无问题，如有必要及时替换。

Stage A: If the temperature drops immediately, check the vacuum bag for leaks. Check equipment and replace if necessary.

B 区：如果固化进程停止，则需拆除真空袋和移除加热设备，重新进行修理。

Stage B: If curing stops, remove vacuum bag and heating equipment and redo repairs.

C 区：如果发生异常，则可继续进行固化直至完成程序，但必须对维修区域进行无损检测，看是否存在扭曲、鼓包、分层、脱胶等缺陷。

Stage C: If an anomaly occurs, curing can proceed until the process is completed; however, non-destructive testing must be performed on the repair area to determine the presence of defects such as distortion, bulging, delamination, and adhesive failure.

D 区：可以接受的修理，按照 SRM 进行检查。

Stage D: Acceptable repairs, inspected in accordance with the SRM.

工单 Work Order

固化任务工单
Curing Task Card

任务名称 Task Topic	固化 Curing
步骤 Steps	规章 / 指令 Regulation/Instruction
注意 Note	注意：室温下自然固化时间较长。加热可以加速固化过程，减少固化时间，具有明显的商业优势。 Note: Curing at room temperature is characterized by an extended duration. Employing heat to expedite the curing process offers notable commercial benefits as it significantly reduces the overall curing time. 注意：最初安装真空袋后需要检查密封质量，确保没有漏气。检查要求：监控真空表，5 min后，真空表上改变的值必须小于5.0英寸汞柱（127毫米汞柱）。 Note：Initially apply a vacuum to the lay up to check the sealing quality of the vacuum bag and to ensure that there are no leaks. The inspection requirements are as follows: monitor the vacuum gauge, and after 5 minutes, the change in value on the vacuum gauge must be less than 5.0 in. (127 mm) of mercury column. 注意：在加热前真空袋密封质量必须符合要求。 Note: This step must be satisfactory before any heat is applied.
1. 放置热电偶 Thermocouple installation	注意：基于质量目的，记录至少三个热电偶的温度。 Note: For quality purposes, record the temperature of at least three thermocouples.
2. 升温 Increase the temperature	使用电热毯 / 烤灯加热，在达到固化温度前，升温速率为1~5 ℉ /min（0.5~3℃ /min）。 Increase the temperature of the heat blanket/heat lamp at a rate between 1 ℉ and 5 ℉ (0.5~3℃) for each minute until it is at the cure temperature. 注意：确保加热毯在修理区域进行均匀加热，温差过大将对修理和相邻区域造成损伤。 Note: Make sure that the heat blanket gives an even temperature around all of the repair. Excessive temperature difference can cause damage to the repair and adjacent structure. 注意：当升温接近需求温度时，树脂与固化剂发生化学反应会迅速地放热。所以，密切关注加热是防止过热的必要措施。 Note: When the increasing temperature approaches the required curing temperature, the chemical reaction occurs between the resin and curing agent, leading to a rapid exothermic reaction. Therefore, closely monitoring the heating process is essential to prevent overheating.
3. 放置绝缘材料 Insulation installation	通常在修理区域温度较低的对侧放置绝缘材料，以保证所有区域在固化循环始终保持在温度范围内；零件中层合板结构域通常温度较低。 Put insulation on the side of the part opposite the repair area that is usually cooler. It will keep all areas of the repair in the cure cycle temperature range. The solid laminate areas of a part are usually cooler.
4. 固化 Curing	设置固化时间并等待固化。Set time and waiting for curing. 注意：不要一次固化十层以上的修理层。使用的最大升温速率为3 ℉ /min（2℃ /min）。如果修理超过10层，必须多次固化。如果违反本规定，将会产生空隙或者鼓包，导致修理效果不能令人满意。 Note: It is not recommended to cure more than 10 layers of repair at one time. The maximum heating rate used should not exceed 3 ℉ (2 ℃) per minute. If more than 10 layers of repair are applied, multiple curing cycles must be used to avoid the risk of voids or blisters that can compromise the quality of the repair.

任务名称 Task Topic	固化 Curing		
步骤 Steps	规章 / 指令 Regulation/Instruction		
5. 降温卸压 Decrease the temperature and release the vacuum pressure	降温最大速率为 5 ℉ /min（3 ℃ /min）。 Decrease the temperature at a maximum rate of 5 ℉ (3 ℃) per minute. 当温度低于 125 ℉（52 ℃）时，卸压。 When the temperature decreases to less than 125 ℉ (52 ℃), release the vacuum pressure. 注意：基于质量控制，记录至少三个热电偶的温度。在固化完成后，卸压时必须将温度降到 50 ℃ (120 ℉) 以下。 Note: For quality purposes, record the temperature of at least three thermocouples. After curing is complete, the release pressure must reduce the temperature to below 50 ℃ (120 ℉). 拆除真空袋。Remove the vacuum bag equipment.		
6. 恢复现场 Restore the site	完成此工序后，要及时关闭设备，清理工具和航材。 To timely close after completion of the process equipment, cleaning tools and materials.		

固化练习题任务工卡
Curing Exercise Task Card

（1）复合材料的固化有哪些不同的方法？讨论它们的优缺点。

What are the different methods used for curing composite materials? Discuss their advantages and disadvantages.

（2）什么是后固化，为什么它在复合材料的固化过程中很重要？后固化是如何进行的？

What is post-cure and why is it important in the curing process of composite materials? How is post-cure performed?

（3）如何优化复合材料的固化过程以达到最高的效率和质量？讨论可能影响固化过程的不同因素以及如何实现质量控制。

How can the curing process of composite materials be optimized for maximum efficiency and quality? Discuss the different factors that can affect the curing process and how they can be controlled.

固化操作工序卡
Curing Operation Log

1. 工具、耗材、设备清点：

Inventory of tools, consumables, and equipment:

2. 辅助材料型号与尺寸记录：

Record of auxiliary material model and dimensions:

3. 热电偶放置数量与位置：

Number and location of thermocouples:

4. 固化温度、压强、时间记录：

Record of curing temperature, pressure and time:

5. 热电偶温度－时间曲线记录：

Record of thermocouple temperature−time curve:

6. 拆袋：

Unbagging:

操作总结：

Summary:

子任务 3　修理后处理
Subtask 3　Post-repair Treatment

【子任务解析 Subtask Analysis】

复合材料构件完成修理后，一般均需采用无损检测技术对修理区域进行质量检查，重点检查其表面质量，有无空隙、分层等影响产品性能的缺陷。若表面质量存在褶皱等缺陷，则需采用砂纸打磨的方式恢复光滑修理表面。若存在严重缺陷，则应清除无效修补材料，重新按流程进行维修。

After the repair of composite components is completed, non-destructive testing techniques are generally required to inspect the repaired area for quality, with a focus on surface quality and the presence of defects such as voids and delamination that can affect product performance. If the surface quality has defects such as wrinkles, the repair surface must be restored to a smooth finish by sanding. If serious defects are found, the ineffective repair material should be removed, and the repair process should be re-performed according to the standard procedure.

通过该任务的训练，学生可掌握修理后质量检查以及表面处理的具体流程。

Through the training of this task, students can master the specific process of post-repair quality inspection and surface treatment.

【子任务分解 Subtask Break-down】

（1）了解修理后处理的定义及内容。

Understand the definition and contents of post repair treatment.

（2）掌握修理质量的检查方法。

Master the inspection methods for repair quality.

（3）了解表面处理的操作流程。

Proficient in the operational procedures involved in surface treatment.

小任务 1 修理后处理的定义是什么，包括哪些内容？

Subtask 1 What is the definition of post repair treatment, and what does it include?

复合材料修理后处理是指在修理完毕后对修理区域进行必要的处理和检验，以确保修理效果符合要求，并延长修理后构件的使用寿命。它包括以下几个方面：

Post-repair treatment of composite materials refers to the necessary processing and inspection of the repair area after repair to ensure that the repair effect meets the requirements and prolong the service life of the repaired component. It includes the following aspects:

（1）修理检验：对修复后的部件进行检验，包括外观检查、尺寸测量、性能测试等；也需通过无损检测技术对修理区域进行检查，以检测表面质量，是否存在空隙、分层等缺陷，保证修理质量符合要求。

Repair inspection: Inspect the repaired parts, including appearance inspection, dimensional measurement, performance testing, etc; Also, use non-destructive testing techniques to inspect the repair area to detect surface quality, the existence of voids, delamination and other defects, to ensure that the repair quality meets the requirements.

（2）表面处理：对修复区域进行表面处理，包括去除多余的树脂、研磨和打磨等，使其表面光滑、平整，符合要求的外观和表面质量。

Surface treatment: Surface treatment of the repaired area, including removing excess resin, grinding and polishing, etc., to make its surface smooth and flat, and meet the required appearance and surface quality.

（3）后续热处理：对于一些高性能复合材料构件，修理后需要进行后续热处理，以消除残留应力、提高材料性能等。

Subsequent heat treatment: For some high-performance composite components, subsequent heat treatment is required after repair to eliminate residual stress and improve material performance, etc.

（4）漆层恢复：对修复区域进行补漆处理，以恢复原有的外观和保护修复区。

Paint layer restoration: Perform paint repair on the repair area to restore the original appearance and protect the repaired area.

（5）文件记录：对修复过程中的所有操作、材料、设备和检验结果进行记录，建立修复档案，以备将来参考和追溯。

Document recording: Record all operations, materials, equipment and inspection results during the repair process, establish a repair file for future reference and traceability.

小任务 2　修理后处理涉及的检测方法和要求有哪些？
Subtask 2　What are the detection methods and requirements involved in post repair treatment?

复合材料修理后检测需要综合使用多种测试设备和技术，对修复区域的结构、性能、化学成分等多方面进行全面的检测和评估，以确保修复后的部位符合飞行适航安全要求。复合材料修理后检测修复区域工作主要包括以下几个方面：

The post-repair inspection of composite materials requires the comprehensive use of multiple testing equipment and techniques to conduct a comprehensive evaluation of the repaired area's structure, properties, and chemical composition, to ensure the repaired area meets flight airworthiness and safety requirements. The main aspects of post-repair inspection of composite materials include the following:

（1）外观检查：通过使用超声波、X 射线、光学投影仪等设备，检测修复区域厚度和形状是否符合要求，以确保修复部位的结构完整性和外观质量。

Visual inspection: Using equipment such as ultrasound, X-ray, and optical projectors to check whether the thickness and shape of the repaired area meet the requirements to ensure the structural integrity and appearance quality of the repaired area.

（2）连接强度：使用拉力试验、压力试验等设备，对修复区域进行负载测试，以评估连接的强度和性能是否满足要求。

Bond strength: Using equipment such as tensile testers, compression testers to conduct load tests on the repaired area to evaluate whether the connection strength and performance meet the requirements.

（3）缺陷和裂纹：使用超声波、红外线热成像等设备，对修复区域进行缺陷检测和裂纹检测，以确保修复后的部位无缺陷和裂纹。

Defects and cracks: Using equipment such as ultrasound, infrared thermal imaging to conduct defect and crack detection on the repaired area to ensure that the repaired area has no defects and cracks.

（4）化学成分：使用光谱仪、质谱仪等设备，对修复区域进行化学成分检测，以确保修复后的部位的材料和化学性质符合要求。

Chemical composition: Using equipment such as spectrometers, mass spectrometers to conduct chemical composition testing on the repaired area to ensure that the material and chemical properties of the repaired area meet the requirements.

（5）热性能：使用热分析仪、热差示扫描仪等设备，对修复区域的热性能进行测试，以确保修复后的部位能够承受高温和低温环境。

Thermal performance: Using equipment such as thermal analyzers, differential scanning calorimeters to test the thermal performance of the repaired area to ensure that the repaired area can withstand high and low-temperature environments.

小任务 3　飞机复合材料的涂料分为哪几类？
Subtask 3　What are the kinds of anti-corrosion surface coating of composite materials?

复合材料涂装是一种重要的表面涂装方式，用于保护复合材料表面免受腐蚀、氧化、化学品侵蚀等的影响，是维护复合材料表面质量和延长使用寿命的重要手段。常见的复合材料涂料包括以下几种：

Composite material anti-corrosion coating is an important surface coating method used to protect composite material surfaces from corrosion, oxidation, chemical erosion and other influences. It is an important means of maintaining surface quality and extending service life of composite materials. Common types of composite material anti-corrosion coating include the following:

（1）底漆涂料：防腐底漆涂装是防腐涂装的基础，主要用于增强表面附着力，提高防腐性和耐久性等。一般使用环氧底漆或聚氨酯底漆进行防腐底漆涂装。

Primer coating: Anti-corrosion primer coating is the foundation of anti-corrosion coating and is mainly used to enhance surface adhesion, improve anti-corrosion and weather resistance. Epoxy primer or polyurethane primer are generally used for anti-corrosion primer coating.

（2）面漆涂料：防腐面漆涂装主要用于保护复合材料表面免受紫外线、化学腐蚀和机械损伤等的影响。一般使用环氧面漆或聚氨酯面漆进行防腐面漆涂装。

Topcoat coating: anti-corrosion topcoat coating is mainly used to protect composite material surfaces from UV, chemical corrosion, and mechanical damage. Epoxy topcoat or polyurethane topcoat are generally used for anti-corrosion topcoat coating.

（3）防腐贴膜涂料：防腐贴膜涂装是一种常用的防腐涂装方式，通过将聚氨酯、聚酰亚胺、氟塑料等防腐材料贴在复合材料表面，形成一层保护膜，能够有效地防止腐蚀和氧化。

Anti-corrosion film coating: Anti-corrosion film coating is a commonly used anti-corrosion coating method. By applying anti-corrosion materials such as polyurethane, polyimide, fluoroplastics to the surface of composite materials, a protective film is formed to effectively prevent corrosion and oxidation.

（4）防腐涂料：在复合材料防腐涂装出现破损、脱落等问题时，可以使用防腐涂料进行修补，以保证涂层的完整性和防腐效果。

Anti-corrosion coating: When problems such as damage or peeling occur in the anti-corrosion coating of composite materials, anti-corrosion coating can be used for repair to ensure the integrity and anti-corrosion effect of the coating.

小任务 4　以 A330 机型为例，飞机复合材料的特殊表面涂料有哪些？
Subtask 4　Taking the A330 aircraft model as an example, what are the special surface coatings for aircraft composite materials?

（1）防腐蚀特殊涂料 Special Coating for Corrosion Protection

防腐蚀特殊涂料旨在防止或延缓腐蚀。在腐蚀保护方面，通常使用一种特殊涂层。这种涂料是由无硅材料制成的挥发性液体，与矿物油有机结合，以排除表面水分。该物质可以通过喷涂或刷涂的方式处理表面。随着液体快速挥发，类似润滑脂的薄膜状可覆盖于机体表面，能够渗透极小空隙并排除水分。

Anti-corrosion special coatings are designed to prevent or delay corrosion. In terms of corrosion protection, a specific type of coating is commonly used. This coating is a volatile liquid made from non-silicon material, combined with mineral oil organics to remove surface moisture. The substance can be applied to the surface through spraying or brushing. As the liquid rapidly evaporates, a thin film similar to lubricating grease forms, covering the surface of the structure, capable of penetrating tiny gaps and excluding moisture.

（2）抗磨损特殊涂料 Special Coating for Abrasion Protection

抗磨损特殊涂料主要使用涂漆（抗磨损涂层），通常应用于容易磨损的部件，如前缘缝翼下的固定前缘。它由聚氨酯树脂（基料和活化剂）组成，通常需要稀释到适合喷涂或刷涂的黏度。

Lacquer (Abrasion resistant coating) is applied to component parts, which are subjected to abrasion, for example, the fixed leading edge of the wing under the slats. It comprises of a polyurethane resin (base and activator), which usually needs to be thinned to the appropriate viscosity for spray or brush application.

（3）抗静电特殊涂料 Special Coatings with Antistatic Properties

涂有导电静电负荷的聚氨酯清漆（抗静电）和结构涂料（抗静电），可将非导电的外部表面的静电负荷传导到飞机金属结构和静电放电器。下面的涂层用于此功能：

Lacquers (antistatic) and structure paint (antistatic) are applied to conduct the electrostatic loads from nonconductive external surfaces to the aircraft metallic structure and static dischargers. The subsequent coatings are used for this function:

1）低电阻抗静电聚氨酯清漆，用于一般结构。

Low resistivity antistatic polyurethane lacquer, used for general structures.

2）高电阻抗静电聚氨酯清漆，用于雷达罩和天线罩。

High resistivity antistatic polyurethane lacquer, used for radomes and antenna fairings.

这些抗静电涂料包括聚氨酯树脂（基料和活化剂），通常需要稀释到适当的黏度，以便喷涂或刷涂。

These antistatic paints comprise of a polyurethane resin (base and activator), which usually needs to be thinned to the appropriate viscosity for spray or brush application.

（4）密封件特殊保护涂层 Special Coating for Protection of Sealant Beads

聚酰胺结构涂料或聚酰胺清漆和柔性聚氨酯涂层结构涂料，此种涂料主要用于密封层可能与液压油接触的区域。密封层不可抵抗酯磷液压油腐蚀，如 Skydrol 液压油。

Structure paint (coating, polyamide) or lacquer (coating, polyamide) and the structure paint (flexible polyurethane coating) are used for their resistance to hydraulic fluids on sealant beads and sealant coatings, which are not self-resistant to ester phosphoric hydraulic fluids like Skydrol.

目前，柔性聚氨酯涂层正在逐渐投入飞机生产设计，以代替聚酰胺结构涂料，其性能更加环保，且与聚酰胺结构涂料完全可互换。

The structure paint (flexible polyurethane coating) is now being introduced on the aircraft in production in place of the structure paint (coating, polyamide). It is more environmentally friendly and are fully interchangeable with Structure paint (coating, polyamide).

修理后处理练习题任务工卡
Post-Repair Treatment Exercise Task Card

（1）请查阅手册，举例说明雷达罩修理后处理涂层的特殊要求。

Please refer to the manual and provide an example of the special requirements for post-repair coating of radar radomes.

（2）请查阅手册，举例说明飞机配平片修理后处理涂层的特殊要求。

Please refer to the manual and provide an example of the special requirements for post-repair coating of aircraft trim tabs.

任务5 封装、固化与修理后处理考核表

姓名		班级		
评价维度	分值／分	自评（30%）	互评（30%）	师评（40%）
素养（20%）				
1.材料清点齐全	5			
2.任务书自学情况	5			
3.安全文明操作及6S	10			
技能（60%）				
1.真空封装湿操作规范性 （封装方式、顺序、辅助材料尺寸、封装真空度）	25			
2.固化方式正确性、操作规范性 （温度、控温速率）	25			
3.个人防护用品穿戴 （防护服、手套）	10			
总结报告（20%）				
1.训练总结的完整性	10			
2.个人反思与拟订后续训练计划	10			
总计	100			
任务完成情况	提前完成			
	准时完成			
	滞后完成			

06 技能综合训练
Task 6
Comprehensive Skills Training

任务6

【任务情境 Task Scenario】

你作为航线技术人员，参加一次紧急维修任务，要求在有限时间内完成复合材料损伤评估和修理任务，需要使用合适的工具和材料，进行损伤类型的识别、定位和修复，同时记录和汇报整个修理过程和结果。

As an airline technician, you participate in an emergency repair task that requires completing composite damage assessment and repair within a limited time using appropriate tools and materials. You must identify, locate and repair damage types while recording and reporting the entire repair process and results.

【任务解析 Task Analysis】

学生应充分应用前期学习的知识与技能要点，独立完成典型的复合材料修理案例，根据任务情境和要求，进行问题分析和解决，从而提高其问题解决能力和创新能力。

Students should apply the knowledge and skill points learned earlier to independently complete typical composite repair cases. They should analyze and solve problems according to the task context and requirements to improve their problem-solving and innovation abilities.

通过本任务的考核，应实现以下目标：

The following goals should be achieved through this task assessment:

知识目标 Knowledge Objectives

（1）了解飞机复合材料的制造工艺和修理方法。

Understand the manufacturing process and repair methods of aircraft composite materials.

（2）了解并掌握行业标准手册 AC43-13、结构维修手册（SRM）。

Know and master the industry standard manuals AC43-13 and Structural Repair Manual (SRM).

（3）了解如何查阅厂家手册，以确定修补用干织物牌号、确定修补用的胶液牌号等完成修理任务所需的重要数据。

Know how to consult manufacturer manuals to determine important data such as the patch fabric and adhesive to be used for repair tasks.

 能力目标 **Ability Objectives**

（1）能够进行飞机复合材料的检测和评估，确定其损伤情况。

Ability to perform inspection and assessment of aircraft composite materials to determine their damage conditions.

（2）能够制定飞机复合材料的修理方案，选择合适的修理方法。

Ability to develop repair plans for aircraft composite materials and select appropriate repair methods.

（3）能够进行飞机复合材料的清洗、干燥、去除损伤和打磨修理型面等操作。

Ability to perform operations such as cleaning, drying, removing damage, and polishing repair surfaces for aircraft composite materials.

（4）能够进行飞机复合材料的湿铺层制备和铺贴操作。

Ability to prepare and lay wet layer of aircraft composite materials.

（5）能够进行飞机复合材料的蜂窝安装和真空袋封装操作。

Ability to perform honeycomb installation and vacuum bag sealing operations for aircraft composite materials.

素质目标 **Emotion Objectives**

（1）具有严谨的工作态度和责任心，认真执行各项操作规程，确保工作质量和安全。

You should have a rigorous work attitude and sense of responsibility, carefully executing all operation procedures to ensure work quality and safety.

（2）增强学生的问题解决能力：训练过程中，学生需要根据任务情境和要求，进行问题分析和解决，从而提高其问题解决能力和创新能力。

Enhance students' problem-solving skills: During the training process, students need to analyze and solve problems based on the task context and requirements, thus improving their problem-solving and innovation abilities.

（3）培养学生的管理能力：针对一些复杂任务，学生需要进行任务计划、资源分配和进度控制等方面的管理工作，从而提高其管理能力。

Develop students' management skills: For some complex tasks, students need to engage in management activities such as task planning, resource allocation, and progress control, thus improving their management skills.

综合训练题一　层合板单面修理

步骤 Steps	指导 / 工作步骤 Instruction / Work steps		
1	用 240 目的砂纸打磨存在缺陷的区域。 阶梯打磨，阶梯搭接宽度为 0.5 in。 额外增加 5 mm 非结构修理层。 Abrade the plies of the affected area to remove the defect with abrasive paper 240 grit. Step sanding with 0.5 in. for each step layup. Sand additional 5 mm of non-structural ply. 注意事项：打磨时不要用压缩空气清除粉尘，应使用吸尘器。 Attention: Do not use compressed air to remove the residue when sanding, use vacuum.		
2	检查损伤，判断打磨层数，以及影响区域的宽度和长度。 Inspect the area, measure number of removed layers, and size of affected area in terms of width and length.		
	尺寸 Size	非结构修理层 Non-structured ply removed	面板打磨层 Glass fiber piles removed
	Conformity Check(Sign to the referee): _____ 一致性检查（示意裁判）：_____		
3	用棉布清洗。勿使用压缩空气进行清洁。 注意事项：立即用干棉布擦拭干净，并风干 10 min。 记录风干时间：开始：____ : ____，结束：____ : ____ Clean the area with a cotton cloth. Do not use compressed air for cleaning. Attention: Wipe the cleaned surfaces immediately with a dry cotton cloth and vent for 10 minutes. Record venting time: Start:____:____, Finish:____:____		
4	湿铺层制备： 需要 1 个填充层 (方向与去掉的第一层相同)。 新层的数量和方向与磨损层一致。 在修补层之上增加一层打磨层（最小搭接宽度为 15 mm）。		

步骤 Steps	指导 / 工作步骤 Instruction / Work steps
4	Lay-up the following plies impregnated: 1 filler ply (orientation as per the first ply removed). New plies with number and orientation as per the abraded plies. One additional cover ply (minimum overlap of 15 mm) above the repair layup. （下表）

识别 Identification	有效期 Expiry Date	数量 Quantity	混合比例 Mixed Ratio
玻纤 Glass Fabric	N/A		
树脂 Resin	N/A		
固化剂 Hardner	N/A		

步骤 Steps	指导 / 工作步骤 Instruction / Work steps
5	按照下图进行封装： 真空袋 Bagging Film 透气毡 Breather Fabric 无孔隔离膜 Non-perforated Film 有孔隔离膜 Perforated Film 可剥离布（用玻纤替代）Peel Ply >40 mm　修理补片 Repair　>40 mm 零件 Part 真空回格 Vacuum Network　真空表 Vacuum Gauge
6	真空读数应达到 22 英寸汞柱。 Vacuum reading should reach at 22 in. Hg
7	在 100℃下固化 70 min。 记录固化参数：开始：＿＿：＿＿，结束：＿＿：＿＿，温度：＿＿℃ Cure for 70 minutes at 100 ℃ . Record cure details: Start:＿＿:＿＿, Finish:＿＿:＿＿, Temperature:＿＿ ℃

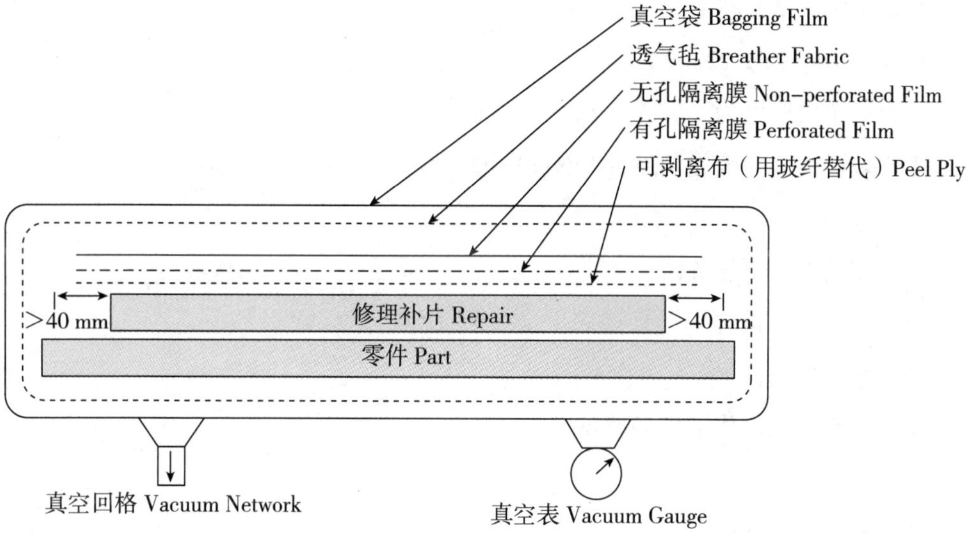

综合训练题二　夹芯结构单面修理

　　某客机航后检查机身，夹层板结构出现凹坑损伤，而且夹层板蜂窝夹芯受损，初步检查得到以下损伤信息：

凹坑尺寸：

长 × 宽 × 深为 1 cm×1 cm×1 cm

凹坑周围存在分层损伤。

零件坐标：

平面 *YOZ*：STA 666；

平面 *XOY*：WL 88；

平面 *XOZ*：RBL 22。

层合板单层厚度为（0.25 ± 0.08）mm。

零件表面前 4 层铺层角度为 [45/0/0 −45]。

选手需完成以下任务：

（1）填写损伤报告；

（2）修复损伤；

（3）详细记录操作工序的要点。

考核时间：210 min。

Post−flight inspection result indicated one dent damage in the sandwich panel structure of the fuselage. The honeycomb core was damaged as well. Preliminary inspection showed the following damage information:

· Dent Size:

length×width×depth = 1 cm×1 cm×1 cm

· Delamination combined with dent.

Damaged sandwich panel coordinate information:

YOZ：STA 666；

XOY : WL 88 ;

XOZ : RBL 22.

· (0.25 ± 0.08) mm thickness for single layer of the surficial laminates.

· Laying−up sequence for surficial laminates is [45/0/0/−45].

Competitors shall fulfill the following tasks:

(1) Fill in the structural damage report;

(2) Repair the damage;

(3) Record key points of operating procedures in detail.

Test period: 210 minutes.

（1）工具、材料清单 Tools and Materials List

分类 Classification	序号 Serial No.	名称 Item	数量 Quantity
工具设备 Tool sand Equipment	1	真空泵 / 个 Vacuum pump/piece	1
	2	真空管 / 个 Vacuum tube/piece	1
	3	真空快速接头 / 个 Vacuum quick coupling/piece	1
	4	真空底座 / 个 Vacuum base/piece	1
	5	烤灯 / 个 Heat lamp/piece	1
	6	电子秤 / 个 Electronic scale/piece	1
	7	原材料存放架 / 个 Storage rack for raw materials/piece	1
	8	清洁工具 / 个 Cleaning tools/piece	3
	9	气动打磨枪 / 个 Pneumatic grinding gun/piece	1
	10	护目镜 / 副 Goggles/pair	1
	11	警示牌 / 个 Warning signs/piece	1

分类 Classification	序号 Serial No.	名称 Item	数量 Quantity
材料 Materials	1	一次性乳胶手套 / 副 Disposable latex gloves/pair	若干
	2	工业防护服 / 套 Industrial protective clothing/suit	1
	3	真空袋薄膜 / 卷 Vacuum bag film/volume	1
	4	聚四氟乙烯布 / 卷 Teflon cloth/volume	1
	5	聚四氟乙烯薄膜 / 卷 Teflon film/volume	1
	6	环氧树脂胶 /kg Epoxy resin/kg	4
	7	固化剂 /kg Curing agent/kg	4
	8	玻纤维布 / 卷 Glass fiber cloth/volume	1
	9	压敏胶带 / 卷 Pressure-sensitive tape/volume	1
	10	复合材料夹层板及芯材 Composite sandwich panel and core material	1
	11	密封腻子条 / 卷 Sealing the putty strip/volume	1
	12	透气毡 / 卷 Breather/volume	1
	13	配胶杯 / 个 Glue cup/piece	若干
	14	注射器（≥ 5 mL）/ 个 Syringe (≥ 5 mL)/piece	1
	15	搅胶棒 / 个 Stir glue stick/piece	1

（2）样图指南

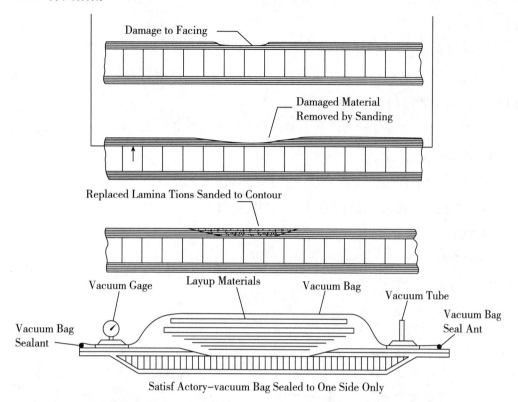

Damage to Facing

Damaged Material Removed by Sanding

Replaced Lamina Tions Sanded to Contour

Vacuum Gage Layup Materials Vacuum Bag Vacuum Tube

Vacuum Bag Seal Ant

Vacuum Bag Sealant

Satisf Actory-vacuum Bag Sealed to One Side Only

Vacuum Gage Layup Materials Vacuum Bag Vacuum Tube

Note：This is Recommended for 350 ℉(177 ℉)Repairs.

Satisfactory-vacuum Bag Sealed Around All of The Part
Parts Which Have Only One Panel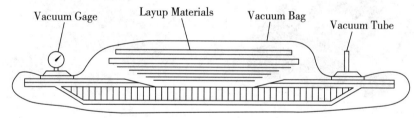

（3）考核程序

① 按照清单清点及检查工具、材料。

② 检查防护用品穿戴是否准确。

③ 结构损伤区域确定：检查损伤部位，确定损伤范围大小、深度等信息，填写结构损伤报告。

④ 修理与除漆区域确定：选择修理方法（斜接法 Scarf Method、阶梯法 Step-Joint Method），确定范围大小、修补层数等，周边贴标记带。

⑤ 打磨移除损伤：用打磨机、砂纸等清除损伤区域。

⑥ 清洁：用蘸有丙酮的干净湿棉布对打磨后的表面进行清洁，检查清洁效果，之后晾干（常温）。

⑦ 湿补片的制作与铺贴：

A. 使用与原始结构相同或厂家允许的玻璃纤维布，并按照修补尺寸与含胶量要求（50±5%），计算胶液用量，并配胶。

B. 将胶液浸入玻璃纤维布，制成湿补片，两面用合适的辅助材料包住，标记方向，并裁剪至所需的尺寸，计算实际含胶量。

C. 按照原铺层方向进行铺贴，清除铺层之间树脂中的空气。

J-4 胶配胶环境要求：温度 15~30 ℃，湿度 ≤ 75%。

J-4 胶配方：（质量比）

环氧树脂胶 100；

邻苯二甲酸二丁酯 5；

三乙烯四胺 10~14。

A. Fill The Glass Cloth With Resin.

B. Cover The Glass Cloth With Peel Ply And Trim It to The Proper Size With Scissors

⑧ 封装与固化：

A. 按照 AC43-13 第 3 章及 SRM51 章第 70 节的要求，进行真空包装，检查真空是否完好（SRM 真空包装样图请自行查阅手册）。

B. 使用设备进行固化，允许选手委托现场裁判或工作人员进行看守。

附 J-4 胶固化参数：常温固化 24 h 以上；加温至 60~100 ℃，固化 2~3 h。

⑨ 清理工作现场，清点工具。

⑩ 结构损伤报告、操作过程记录及后续工序填写。

（4）比赛所用资料

① 行业标准规范手册：AC43.13-1B CHAPTER 3. FIBERGLASS AND PLASTICS；

② 厂家结构修理手册 (参考 Boeing737-800)：SRM 51-70-04 REPAIR PROCEDURES FOR WET LAYUP MATERIALS。

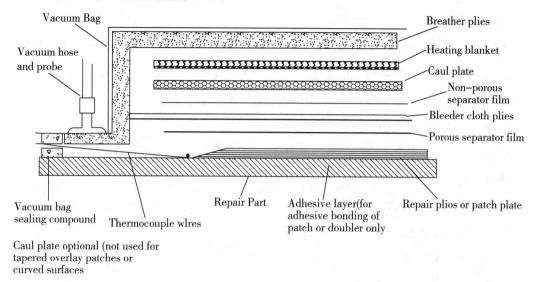

Vacuum Bag

Breather plies

Heating blanket

Caul plate

Non-porous separator film

Bleeder cloth plies

Porous separator film

Vacuum hose and probe

Vacuum bag sealing compound

Thermocouple wlres

Repair Part

Adhesive layer(for adhesive bonding of patch or doubler only

Repair plios or patch plate

Caul plate optional (not used for tapered overlay patches or curved surfaces

综合训练题三　B737 升降舵修理

（1）任务描述

该零件为一架 B737-800 飞机的升降舵，中国制造，登记标志号为 2641，飞机序列号为 28056，飞行总时长为 4 150 h，经历 2 450 个飞行周期，上次大修日期为 2019.1.30。根据 MPD 需对此零件进行检修，提交损伤报告及维修过程记录表。

发布人：×××，发布日期：2023.4.14

附：

零件坐标：

平面 YOZ：STA 1217；

平面 XOY：WL 208；

平面 XOZ：RBL 2。

注意：默认单位为英寸（in）。

零件表面前 3 层铺层角度为 [±45°/0/90/±45°]。

（2）样图指南

图线名称	图线形式	图线宽度	一般应用举例
粗实线	——————————	d（粗）	可见轮廓线
细实线	——————————	$d/2$（粗）	尺寸线及尺寸界线 剖面线 重合断面的轮廓线 过渡线
细虚线	— — — — — —	$d/2$（细）	不可见轮廓线
细点画线	— · — · — · —	$d/2$（细）	轴线 对称中心线
粗点画线	— · — · — · —	d（粗）	限定范围表示线
细双点画线	— ·· — ·· — ·· —	$d/2$（细）	相邻辅助零件的轮廓线 轨迹线 极限位置的轮廓线 中断线
波浪线	∿∿∿	$d/2$（粗）	断裂处的边界线 视图与剖视的分界线
双折线	─/\─/\─	$d/2$（粗）	同波浪线
粗虚线	▬ ▬ ▬ ▬ ▬	d（粗）	允许表面处理的表示线

名称	符号	说明及应用	图例
中心线	$\underset{L}{\mathcal{C}}$	用于表达物体的中心轴或对称中心平面	STRUT WL97 $\underset{L}{\mathcal{C}}$ ENGINE
旗标		在旗标符号内标注数字、字母或符号用于表达旗标箭头所指处的标记，其详细说明在零件清单中描述 注：字母或符号仅在特殊场合应用	NAG 1304–150 NAG 43DD4–19(2) AN 960 D416 AN 310–4 MS 24665–153 〔1〕 〔1〕 NSTALL COTIER PIN PER BAC 5018
方向指示	UP INBD ←FWD	表明视图或者某零构件相对于飞机坐标的方位	VIEW A–A UP FWD→
站位	STA 360	用于表示机身站位（STA）、水线站位（WL）、纵剖线站位（BL）	STA 360 WL 92.74

（3）考核步骤

1）操作前清点检查记录：

①工具检查：例如，工具清单完整，定检标识完整，均在有效期内。

②材料检查：例如，材料清单完整，合格证明完整，发现固化剂超期，其余均在有效期内。

2）结构损伤区域确定记录：

①目视检查依据：_____ 手册_____ 章节号，_____ 段_____ 页，进行操作。经过目视检查，发现损伤有以下几种：_____、_____、_____，即需包括损伤以下信息（位置）（数量）（种类）（备注：定性按照损伤报告表中的分类来选择即可）。

②敲击检测：

a.查询构件图纸，获得构件的内部结构信息（铺层层数、方向、厚度等）。

b.在工件表面上标注内部结构和检测区域，画出示意图。

c.敲击检测：按照××手册××章节号，××段××页。其敲击检测结果如下：

损伤面积____，长____，宽____。根据手册要求，损伤面积超过____，可认为其为可修理损伤。

d.对于划伤、凿伤、凹坑、蜂窝压陷等类型损伤，进行规范打磨确认损伤深度。____缺陷经打磨确认损伤层数为____，根据单层标准厚度理论值____，板厚公差为____，则损伤范围深度值为____（精确到小数点后2位）。

e.打磨移除损伤后，测量蜂窝损伤深度约为____。

f.损伤打磨区域用胶带进行保护，打磨完成后进行清洁。

g.填写损伤记录表和相关文件（备注：内部损伤的深度信息需采用其他检查手段进行确定）。

3）修理与除漆区域确定记录：

①确定打磨方法：按照____手册____章节号，____段____页，确定维修方法为阶梯打磨。

②理论阶梯打磨数据：损伤层数为____，损伤大小为____。最大修理区域（含/不含气动打磨层）____。因此各层尺寸为Ply1：大小____，角度____，Ply2：大小____，角度____，气动打磨层（若有）：大小____，角度____。

③按____手册____章节号，____段____页，确定除漆区域比最大修理区域半径大____，区域外胶带保护。

4）打磨移除损伤记录：

①构件除湿：请示后，得到____指令并记录。

②记录蜂窝镂铣的面积____，并圈定蜂窝镂铣的外围圆，____保护。

③镂铣蜂窝：记录镂铣蜂窝的大小____，深度____。

④阶梯打磨：画图记录每层的角度、大小，画出实际打磨尺寸示意图。

5）清洁记录（粘接型面）：

①用清洁剂____和无毛布（干、湿擦布协同）将粘接面清洁。

②晾干（因现场有灰尘，可能需要请示裁判，记录晾干时长）。

③保护粘接面（裁剪薄膜覆盖）。

6）湿补片的制作与铺贴记录：

①修补材料牌号为____，记录称重织物质量____g。

②记录温度____，湿度____。若超标，记录当前值____，请示裁判，得到许可后进行操作。

③根据____手册/文件，理论胶液质量为____。实际胶液质量为____，按顺序配置环氧树脂为____g，邻苯二甲酸二丁酯为____g，三乙烯四胺为____g。符合配置比例____。搅拌时长为____min，静置时间为____min。

④按____手册规范制备湿铺层，记录湿铺层方向、辅助材料及玻纤布的尺寸关

系：____。

　　⑤ 裁剪补片：实际补片大小和角度为 Ply 1: 大小____，角度____，Ply 2: 大小____，角度____，气动打磨层（若有）：大小____，角度____。

　　⑥ 实际含胶量为____，符合技术手册/文件中含胶量要求。

　　7）蜂窝芯准备：

　　① 裁剪蜂窝芯高度为____，尺寸为____。

　　② 清洁：在____溶剂中浸泡____ s（不超过 60 s）。

　　③ 干燥：采用____（吹风枪）将蜂窝芯干燥完全。

　　④ 保护蜂窝芯（裁剪薄膜覆盖）。

　　8）蜂窝芯安装与铺叠：

　　① 按照____手册安装蜂窝芯。

　　② 按照____手册铺叠裁剪好的补片。

　　③ 铺叠补片过程中进行排气。

　　9）封装与固化记录：

　　① 封装方式为____（单面/双面真空封装）。

　　② 按照____手册，____章节号，____段____页，其实际辅助材料的顺序、尺寸大小，画封装示意图：____。

　　③ 真空度为____ MPa，满足____手册对固化压力的要求。

　　④ 固化温度为____，固化时间为____。固化过程已指定看守。

　　10）现场清洁、工具清点记录：

　　① 清点工具：有无缺少，定检标识是否有缺少或污染等。

　　② 清点材料：若有用完的，需记录。

　　③ 清洁记录：个人桌面、共用桌面、打磨台、地面、材料架。

　　④ 清洁工具：个人和共用的工具（电子秤等），用溶剂擦拭清洁。

　　11）后续方案的填写：

　　① 拆袋目视检查表面质量。

　　② 视情况光滑修理表面。

　　③ 使用敲击或其他检测方法检验修理质量，若不合格，重新进行修理；若合格，则进行下步操作。

　　④ 恢复涂层。

　　⑤ 完成配平、配重等其他要求。

记录人：

记录时间：

TEST PROJECT AIRCRAFT MAINTENANCE

Competitor's Working Document

NAME	(First)		(Last)
COUNTRY			
START TIME			

Objective

To test the Competitors ability to carry out composite inspection and repair in accordance with manufacturer's recommendations and company procedures.

Time Allotted

4 hours

Process

PART 1

Carry out **NDT Inspection** in accordance with company procedures. Details of the affected system will be provided in the form of a verbal briefing by your supervisor. Record findings on the **Structural Damage Report**. Only defects affecting airworthiness will be considered. Submit Structural Damage Report and **Maintenance Planning**.

PART 2

Correct handling of honeycomb panel in accordance with aircraft manuals. Composite repai rosette fitting completed to within the Manufacturer's Airworthy Tolerance. Submit **Maintenance Log Recording**.

Notes

Do not correct any defects or findings during the inspections.

The "**Judge**" and "**Supervisor**" is the same individual with respects to this module.

The backside of the Contestant Document can be used for notes and is not subject to evaluation.

Reference Documents

(1) STRUCTURE REPAIR MANUAL

B737-800 51-70-00 Repair Procedure-General Data;

A350-XWB5 51-73-00 COMPOSITE REPAIR-COMMON DATA;

B787-9 51-73-00 COMPOSITE REPAIR-COMMON DATA;

B787-9 51-23-01 Absorbed Moisture Removal-Heat Blanket -Procedure- General Data

(2) AC-43-13.b

CHAPTER 3 SECTION 1 FIBREGLASS and PLASTICS;

CHAPTER 5 NDI

(3) WS Aviation Maintenance Policy Manual

(4) Independent Control Check Authorization Record

Evaluation Scheme

SECTION-Composite inspection and repair	Relative Importance (%)
PART 1-Working Procedure for Composite Inspection	30
PART 2-Working Procedure for Composite Repair	35
PART 3-Paper Work	30
Airmanship-PPE Selection & Usage, Aircraft Handling, Tool Selection & Usage, Tool Control, Work Flow, Discipline, Cleanliness and Organization	5
Total	100

Scenario

You are an aircraft maintenance employee of WS Aviation who hold a composite part repair certification authority. You arrive at work in the morning ready to begin your shift. Your supervisor informs you that the night shift has carried out a series of inspection tasks. *Post-flight inspected two dents. These two dents are on the right side of the fuselage, between FR 24-25, right stringer 12-13*. Preliminary inspection result will be mentioned in **Maintenance Planning Data**. Your job is to carry out a detailed Inspection, submit the Maintenance Planning and then accomplish the repair task.

Obtain a briefing from your supervisor as to the tasks carried out by the night shift and the system requiring **Maintenance Planning Data**.

All verbal briefing will be available from you supervisor at any point and can be repeated upon request without penalty.

Briefing Notes (no evaluation)

Maintenance Planning Data

Post−flight inspected two dents.

These two dents are on the right side of the fuselage:

between FR 24−25, right stringer 12−13.

Preliminary inspection indicated the following information:

· Dent I

Length×width×depth = 26 mm×13 mm×2 mm;

FR24 spacing distance 10 cm;

Right STR12 spacing distance 15 cm.

· Dent II

Length×width×depth = 10 mm×8 mm×1 mm;

FR25 spacing distance 18 cm;

Right STR13 spacing distance 14 cm;

· One repaired area

6 cm from Dent I

Length×width×depth = 12 mm×12 mm×3 mm;

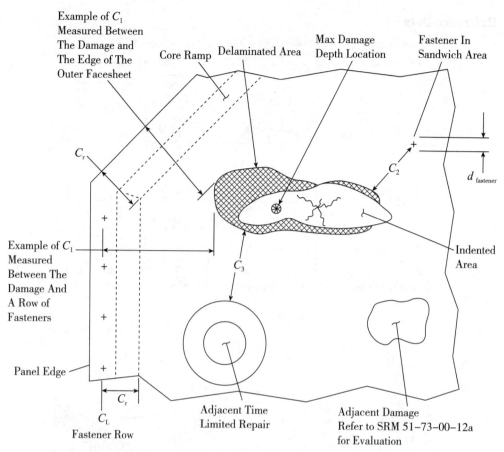

Example of C_1 Measured Between The Damage and The Edge of The Outer Facesheet

Core Ramp

Delaminated Area

Max Damage Depth Location

Fastener In Sandwich Area

C_r

C_2

$d_{fastener}$

Example of C_1 Measured Between The Damage And A Row of Fasteners

C_3

Indented Area

Panel Edge

C_r

C_L

Fastener Row

Adjacent Time Limited Repair

Adjacent Damage Refer to SRM 51-73-00-12a for Evaluation

C_1 = Minimum Distance From The Edge of The Damage to A Fastener Row Centerline (Or to A Panel Edge, If There Are No Fasteners Along That Edge).

C_2 = Minimum Distance From The Edge of The Damage to The Edge of A Fastener Located In The Sandwich Area of The Panel.

C_3 = Minimum Distance From The Edge of The Damage to The Edge of A Repair. This Distance is Also Applicable When The Damage and Repair Are Not on The Same Facesheet.

C_r = Distance From The Top of The Core Ramp to A Fastener Row Centerline or to A Panel Edge, If There Are No Fasteners Along That Edge. C_r Value Given in Srm ADL Section, As Applicable.

Reference Data

Single–Site Dent ADLss Curve for Zone A

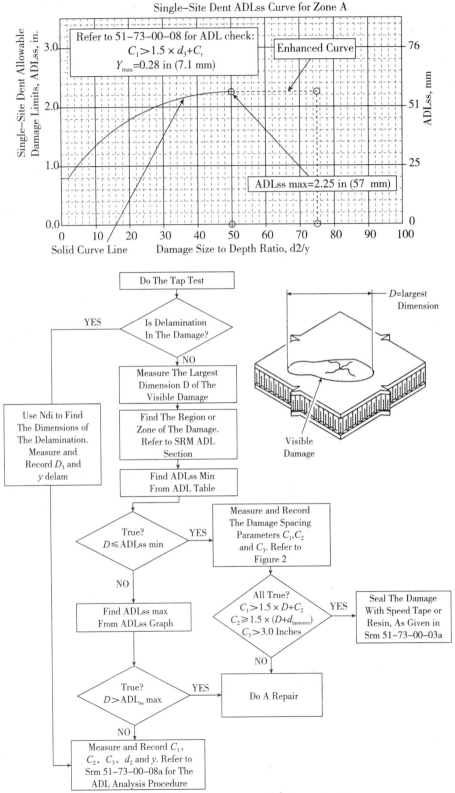

Refer to 51–73–00–08 for ADL check:
$$C_1 > 1.5 \times d_3 + C_r$$
$$Y_{max} = 0.28 \text{ in (7.1 mm)}$$

Enhanced Curve

ADLss max=2.25 in (57 mm)

Solid Curve Line

Damage Size to Depth Ratio, d2/y

Single–Site Dent Allowable Damage Limits, ADLss, in.

ADLss, mm

Do The Tap Test

Is Delamination In The Damage? — YES

Use Ndi to Find The Dimensions of The Delamination. Measure and Record D_3 and y delam

NO

Measure The Largest Dimension D of The Visible Damage

Find The Region or Zone of The Damage. Refer to SRM ADL Section

Find ADLss Min From ADL Table

True? $D \leq$ ADLss min — YES

NO

Find ADLss max From ADLss Graph

True? $D >$ ADL$_{ss}$ max — YES

NO

Measure and Record C_1, C_2, C_3, d_2 and y. Refer to Srm 51–73–00–08a for The ADL Analysis Procedure

Measure and Record The Damage Spacing Parameters C_1, C_2 and C_3. Refer to Figure 2

All True? $C_1 > 1.5 \times D + C_2$ $C_2 \geq 1.5 \times (D + d_{fastener})$ $C_3 > 3.0$ Inches — YES

NO

Seal The Damage With Speed Tape or Resin, As Given in Srm 51–73–00–03a

Do A Repair

D=largest Dimension

Visible Damage

Flow Chart of The Primary Assessment of An Impact Single Damage

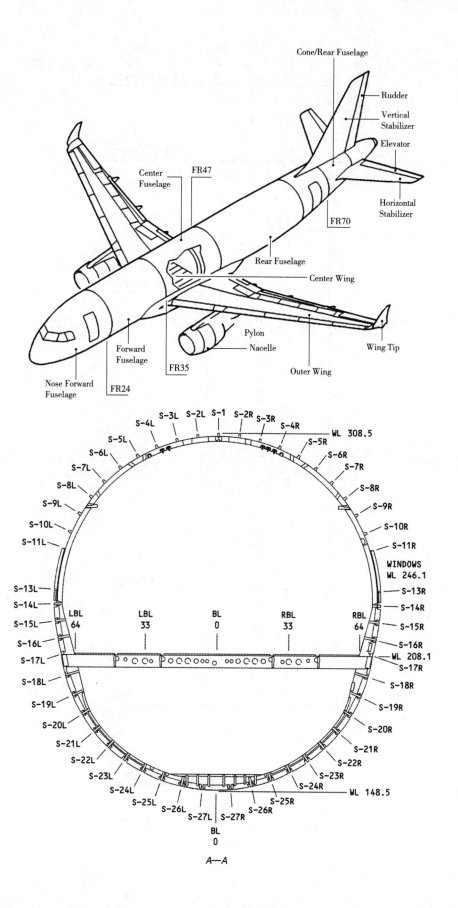

A—A

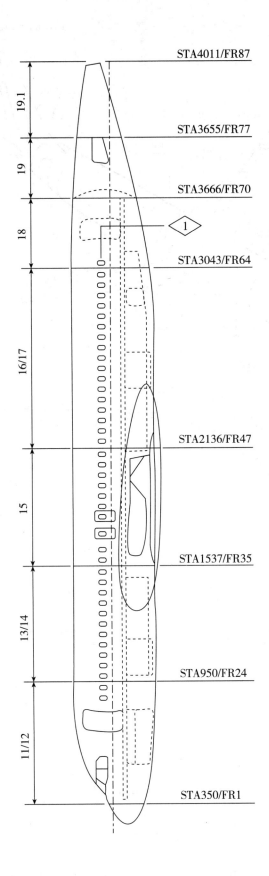

TASK CARD

(1) Check all the tools and material iaw the Tool List & Consumable Material List.

(2) Carry out inspection. Fill in the SDR. Submit MPD.

(3) Remove the damaged skin and core and prepare the repair area.

(4) Follow Step–joint method iaw Fig.1. Lightly abrade the original adhesive and remove remaining core on the bottom of the hole. Use abrasive cloth 240 grade to remove the adhesive fillet and to get a smooth matt surface. Be careful not to remove all of the bonding adhesive. Record repair plies parameters in Maintenance Log Recording (MLR).

(5) Clean the area with a vacuum cleaner.

(6) Clean the inner bonding surface with cleaning agent.

(7) Dry the component iaw Fig.2.

(8) Prepare the honeycomb core plug. Record core parameters in MLR.

(9) Prepare the adhesive paste iaw Fig.3. Record repair patch layout in MLR. The adhesive compound is made up with resin to hardener ratio of 3. Record component of adhesive compound in MLR.

(10) Impregnate the adhesive paste iaw Fig.4.

(11) Abrade or cut the top surface of honeycomb core plug to the same level with the skin surface.

(12) Clean the repair area with cleaning agent.

(13) Install the core plug iaw Fig.5.

(14) Apply adhesive paste to the mating surface Fig.6. Remove all unwanted adhesives. Calculate the fiber ratio and record in MLR.

(15) Put the parting film on the repair area and let cure under pressure. To apply pressure, use a vacuum bag iaw Fig.7. Record the pressure and temperature in MLR.

(16) Remove the equipment and parting film.

(17) Inspect the adhesive fillet and tap test inspect the repair area.

(18) Do not remove the adhesive fillet and seal the edge of the repair with sealant.

(19) Restore the surface protection.

(20) Clean the work area and tools.

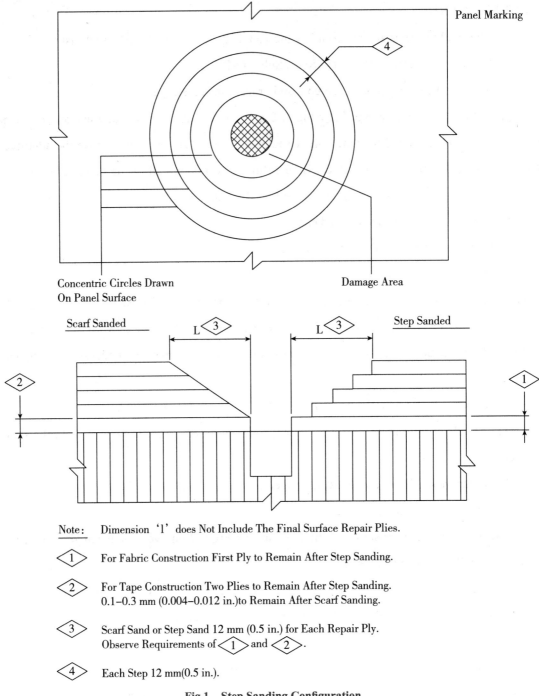

Panel Marking

Concentric Circles Drawn
On Panel Surface

Damage Area

Scarf Sanded   L $\langle3\rangle$ L $\langle3\rangle$ Step Sanded

Note： Dimension '1' does Not Include The Final Surface Repair Plies.

$\langle1\rangle$ For Fabric Construction First Ply to Remain After Step Sanding.

$\langle2\rangle$ For Tape Construction Two Plies to Remain After Step Sanding.
0.1–0.3 mm (0.004–0.012 in.)to Remain After Scarf Sanding.

$\langle3\rangle$ Scarf Sand or Step Sand 12 mm (0.5 in.) for Each Repair Ply.
Observe Requirements of $\langle1\rangle$ and $\langle2\rangle$.

$\langle4\rangle$ Each Step 12 mm(0.5 in.).

Fig.1 Step Sanding Configuration

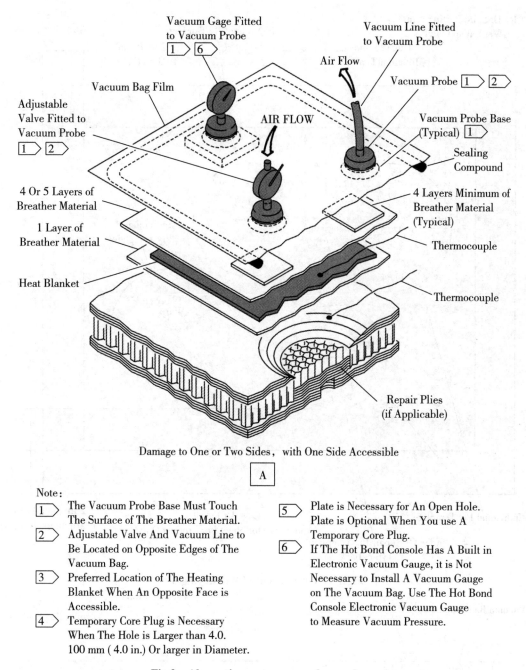

Vacuum Gage Fitted to Vacuum Probe ①▷⑥▷

Vacuum Line Fitted to Vacuum Probe

Air Flow

Vacuum Bag Film

AIR FLOW

Vacuum Probe ①▷②▷

Adjustable Valve Fitted to Vacuum Probe ①▷②▷

Vacuum Probe Base (Typical) ①▷

Sealing Compound

4 Or 5 Layers of Breather Material

4 Layers Minimum of Breather Material (Typical)

1 Layer of Breather Material

Thermocouple

Heat Blanket

Thermocouple

Repair Plies (if Applicable)

Damage to One or Two Sides, with One Side Accessible

A

Note:

①▷ The Vacuum Probe Base Must Touch The Surface of The Breather Material.

②▷ Adjustable Valve And Vacuum Line to Be Located on Opposite Edges of The Vacuum Bag.

③▷ Preferred Location of The Heating Blanket When An Opposite Face is Accessible.

④▷ Temporary Core Plug is Necessary When The Hole is Larger than 4.0. 100 mm (4.0 in.) Or larger in Diameter.

⑤▷ Plate is Necessary for An Open Hole. Plate is Optional When You use A Temporary Core Plug.

⑥▷ If The Hot Bond Console Has A Built in Electronic Vacuum Gauge, it is Not Necessary to Install A Vacuum Gauge on The Vacuum Bag. Use The Hot Bond Console Electronic Vacuum Gauge to Measure Vacuum Pressure.

Fig.2　Absorption water removal procedure

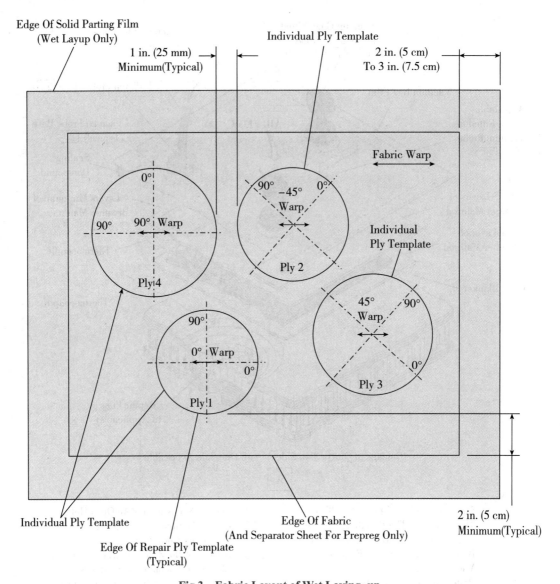

Fig.3   Fabric Layout of Wet Laying-up

Fig.4   Repair Patch Impregnation Procedure

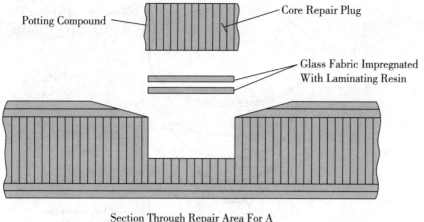

Section Through Repair Area For A
Full Depth Core Replacement

C

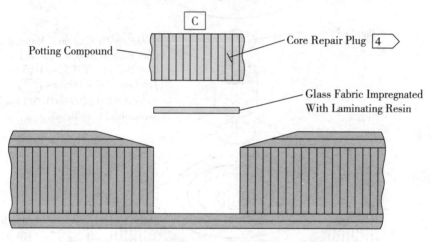

Section Through Repair Area for A
Full Depth Core Replacement

B

Fig.5　Layout of core installation

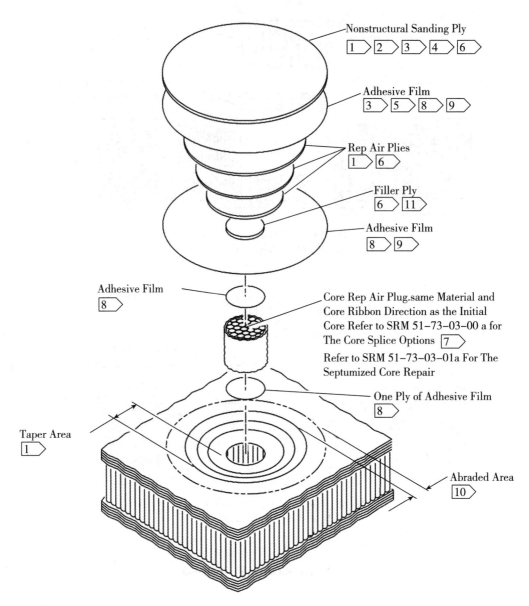

Nonstructural Sanding Ply
1 2 3 4 6

Adhesive Film
3 5 8 9

Rep Air Plies
1 6

Filler Ply
6 11

Adhesive Film
8 9

Adhesive Film
8

Core Rep Air Plug.same Material and
Core Ribbon Direction as the Initial
Core Refer to SRM 51–73–03–00 a for
The Core Splice Options 7
Refer to SRM 51–73–03–01a For The
Septumized Core Repair

One Ply of Adhesive Film
8

Taper Area
1

Abraded Area
10

Fig.6   Layout of Repair Patch Installation

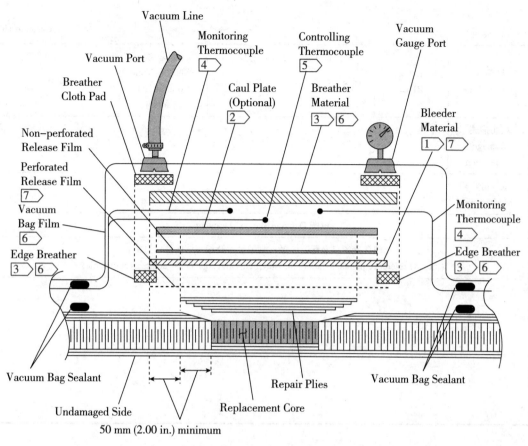

Fig.7   Layout of Vacuum Bagging System structural damage report

Structural Damage Report

损伤 类型 Damage Type	裂纹 Crack ☐		凹坑 Dent ☐		刺穿 Puncture ☐		分层 / 脱粘 / 空谷 Delamination/Debond/Void ☐	腐蚀 Corrosion ☐
	断裂 Rupture ☐		屈曲 Buckle ☐		划伤、缺口、凿伤 Scratch/Nick/Gouge ☐		其他 Other _____ ☐	
Exact Damage Position 准确的损伤位置			Damage Dimension 损伤尺寸			Main Damaged Part 主要损伤零件		
站位 Station	mm		长度 Length	mm		件号 Part #		
水线 Water Line	mm		宽度 Width	mm		序列号 Serial #		

185

纵剖线 Buttock Line	LH☐ RH☐	深度 Depth	mm	距离上一次大修的时间 Time from the last overhaul	hours
损伤描述 Damage Description					
成因分析 Mechanism					
草图 Damage Sketch / Drawing					
（包含参考点坐标、位置、测量尺寸及相邻结构件信息） (Include reference points, location, measurements and adjacent structure as applicable)					

Maintenance Log Recording

_____

_____

_____

_____

_____

_____

_____

_____

_____

_____

_____

_____

_____

任务 6 综合技能训练考核表

序号	考核要素	考核要求	评分标准	分值/分
1		操作前工具清点检查	错误、遗漏一处扣 0.5 分，直至扣完分	1
	1.1	检查是否缺少工具或对应配件		0.5
	1.2	检查工具是否有定检标识，判断是否在定检期		0.5
2		操作前材料清点检查	错误、遗漏一处扣 0.5 分，直至扣完分	2
	2.1	检查是否缺少材料		0.5
	2.2	检查材料是否有合格证、出入库情况，判断是否在保质期		1.5
3		结构损伤区域确定	错误、遗漏一处扣 1 分，直至扣完分	15
	3.1	目视检查确定表面损伤情况	包含损伤类型、长×宽、形状坐标系内位置、目视检查的动作	5
	3.2	敲击检查确定是否有内部损伤与判断损伤区域（形状与面积）	发现有内部损伤得 1 分 判断损伤位置正确得 1 分、判断形状正确得 1 分、判断内部损伤区域与实际损伤面积大小公差 ±7.5% 得 2 分，±15% 得 1 分，±20% 得 0.5 分，≥20% 不得分 未发现内部损伤，此项不得分	5
	3.3	确定损伤深度及纤维损伤层数	损伤深度的判定超出实际损伤的深度公差范围	2
		注：假设单层纤维厚度理论值为 0.2 mm，因板材厚度公差为 8 %，若发现损伤 3 层，则损伤深度范围值为 0.2 mm×3×（1±0.08）≈ 0.55~0.65 mm。测量数据要求在公差范围内精确到小数点后 2 位		
	3.4	个人防护用品穿戴（防护服、口罩/面具、手套、护目镜）	遗漏一项扣 0.25 分，直至扣完分	1
	3.5	记录目视检查、敲击检测的参考资料依据	注意：参考资料必须详细描述	2
4		修理与除漆区域确定		6
	4.1	按相关手册确定合适的修理方法（斜接法、阶梯法）		3
	4.2	依据损伤层数正确确定修理区域及尺寸（阶梯法满足 1/2~1 in，斜接法 30 T~50 T）	尺寸错误一处扣 1 分	2
	4.3	正确确定除漆区域及尺寸	尺寸错误扣 1 分	1

序号	考核要素	考核要求	评分标准	分值 /分
5		打磨移除损伤		7.25
	5.1	检查打磨后各尺寸（每层圆直径及阶梯宽度等），均要满足一般公差（±1 mm），检测点不少于2处	尺寸超差一处扣0.5分，直至扣完分	2
	5.2	打磨深度符合要求	多打磨一层扣1分，多打磨两层和伤及蜂窝均不得分	2
	5.3	打磨区域周边保护		1
	5.4	个人防护用品穿戴（防护服、防尘口罩/面具、手套、防尘护目镜、耳塞）	遗漏一项扣0.25分，直至扣完分	1.25
	5.5	打磨后及时关闭设备		1
6		清洁	错误一处扣分，直至扣完分	3.75
	6.1	对打磨后表面进行清洁至擦拭无灰尘		1
	6.2	晾干		1
	6.3	晾干后对清洁表面进行保护		1
	6.4	个人防护用品穿戴（防护服、防毒口罩/面具、手套）	遗漏一项扣0.25分，直至扣完分	0.75
7		湿铺层的制作与铺贴	错误、遗漏一处扣0.5分，直至扣完分	10
	7.1	选择修补材料牌号，并裁剪称重		0.5
	7.2	计算胶液用量		0.5
	7.3	根据环境温湿度，判断是否可以开始配胶		0.5
	7.4	配胶顺序（先放环氧树脂，最后放固化剂）		0.75
	7.5	正确配比胶液		0.5
	7.6	搅拌	不少于2 min	1
	7.7	试剂取用操作（注射器禁止混用）		0.5
	7.8	垃圾分类处理，危废品单独处理，并做标识		0.5
	7.9	刷胶操作（沿纤维方向），至浸透玻璃布		0.5

序号	考核要素	考核要求	评分标准	分值/分
7	7.10	刷胶结束后，涂胶面不得有多余物		0.5
	7.11	刮板排除湿铺层内部气泡		0.5
	7.12	剪出与修补区域大小匹配的补片		0.5
	7.13	含胶量计算	含胶量控制在 50%±5% 内，若超过公差范围，该项不得分	1
	7.14	铺贴区域周边防护		0.5
	7.15	铺贴补片，并排气		0.5
	7.16	清理铺贴区周边余胶		0.5
	7.17	个人防护用品穿戴（防护服、防毒口罩/面具、手套）	遗漏一项扣 0.25 分，直至扣完分	0.75
8		真空包装与固化		7.5
	8.1	确定真空包装方式（单面、双面）		1
	8.2	辅助材料铺放顺序符合标准	错误一处扣 0.5 分	2
	8.3	辅助材料的尺寸符合修理手册要求	尺寸错误一处扣 0.5 分	2
	8.4	真空气密性良好（无明显漏气声）		1
	8.5	确定固化方式 若需现场人员看守设备，请与现场裁判请示		1
	8.6	个人防护用品穿戴（防护服、手套）	遗漏一项扣 0.25 分，直至扣完分	0.5
9		现场清洁、工具清点		4.5
	9.1	工具清洁		1
	9.2	工作台面、电子秤及设备台面清洁		1.5
	9.3	工具清点		1
	9.4	材料清点		1
10		固化后零件质量（是否缺胶气泡、敲击检测）		2
11		入场即展示选手牌		0.5
		取下选手牌前请向裁判请示		0.5

序号	考核要素	考核要求	评分标准	分值/分
12		结构损伤报告单填写	漏填、错填一处扣1分，直至扣完20分	20
	12.1	规范填写结构损伤报告单，参见 WS Aviation Form 8.8–Structural Damage Report		10
	12.2	损伤简图描述（标注内容信息规范、完整、简洁）		10
13		修理工序记录及后续方案（后续方案参考波音 737 SRM）	遗漏、错误、不翔实一处扣1分，直至扣完20分	20
	13.1	操作前清点检查记录		4
	13.2	结构损伤区域确定记录		2
	13.3	修理与除漆区域确定记录		2
	13.4	打磨移除损伤记录		1
	13.5	清洁记录		1
	13.6	湿补片的制作与铺贴记录		4.5
	13.7	封装与固化记录		2
	13.8	现场清洁、工具清点记录		1
	13.9	后续方案的填写		2.5
14		合计		100

参考文献
References

［1］虞浩清，刘爱平.飞机复合材料结构修理［M］.北京：中国民航出版社，2010.

［2］吴悦梅，付成龙.树脂基复合材料成型工艺［M］.西安：西北工业大学出版社，2020.

［3］波音公司.BOEING 787飞机维修手册.

［4］姜波.飞机检测与维修实用手册［M］.长春：吉林科学技术出版社，2005.

［5］Dorworth LC, Gardiner GL, Mellema GM. Essentials of advanced composite fabrication and repair［M］. Newcastle, Washington: Aviation Supplies & Academics, Inc, 2010.